CAVES

CAVES

EXPLORING HIDDEN REALMS

MICHAEL RAY TAYLOR

NATIONAL GEOGRAPHIC

WASHINGTON, D.C.

CONTENTS

Left: A French caver rappels into Chromosome X, a short-lived vertical shaft within a glacial ice cave in Greenland. Top: Janot Lamberton clears loose ice from a Greenland pit. Middle: Pulled along by an electric scooter, a cave diver glides past ancient formations in a Yucatán cave. Bottom: Caver-scientist Hazel Barton admires pristine calcite formations of Fairy Cave in Glenwood Springs, Colorado.

I met Michael Ray Taylor during a caving expedition to China in 1988. We were among the first Americans invited to explore caves in the mountainous limestone regions of southern China, home of the misty, mysterious karst towers depicted in oriental art. Mike and I got to know each other during trips to places like Tontianlou or "the basket of heaven"—a bowl-shaped vertical cave 300 feet deep and so wide that a small forest grew in its bottom. Together we mapped and photographed roaring underground rivers and stalactite-draped chambers never before seen by human eyes. Each night we endured long banquets, trading toasts of bitter rice liquor with our Chinese hosts, only to rise early the next morning and head for the caves.

When Mike sent me this book, I thought back to the beginnings of my own caving career, to our time in China, and to some of the controversies that brewed in the caving community over the dangers of publicity in films and books—fears that cavers who share their secrets with the world might ultimately bring harm to the places they most want to protect. Yet as I read through the text, I became lost again in the wonder of the caves themselves, lost in the adventures experienced by Nancy Aulenbach and Dr. Hazel Barton, two cavers featured in MacGillivray Freeman's IMAX® theater film, *Journey Into Amazing Caves*. I felt the coolness of rock around me, felt the air moving toward a cave's entrance, heard the whisper of Earth's voice—and I found answers to simple yet profound questions: Why are Earth's subterranean lands and waters important? How do caves below relate to life above?

Much of the world's fresh water—about 25 percent in the United States—is stored as groundwater in caves and karst, or limestone landscapes. The protection and management of these vital water resources are critical to public health and to sustainable economic development. As identified by the National Geographic Society, water resources are a critical concern as mankind enters the 21st century. Serving as storehouses of information on natural environments, human history, and evolution, caves contain valuable data relevant to global climate change, waste disposal, groundwater supply and contamination, petroleum recovery, and, as Mike discusses, biomedical investigations. Caves also contain vast resources important to the sciences of anthropology, archaeology, geology, paleontology, and mineralogy.

The U.S. government considers caves such valuable natural resources that in 1988 a federal law was passed to protect caves and all that they contain. This law was the result of the time and efforts of cavers and conservationists throughout the nation; it is one of the few acts ever passed by Congress to protect a specific natural resource. And yet many people, in fact the great majority, still think of caves negatively and will only enter passages that have been altered and harnessed with asphalt trails, electric lights, elevators, and containing, in the saddest cases, underground lunchrooms. While managed tours may be a good approach for the protection of some caves from destruction by vandals and inexperienced cavers, they do not adequately inform people about the interconnectedness of the living planet and its life-forms—that the vast cave networks of the world are not simply curiosities but important components of the Earth's natural systems.

Here then, finally in these pages, are the force and magic of words and pictures that offer a connection between two worlds—above, the world of sunlight, blue sky, white clouds, and birdsong; below, a world of darkness, filled with the roar of water rushing between icy walls, of great chasms dropping to unplumbed depths, and of water bursting from fountains not built by the hand of man. The connection is made with Mike's informative and entertaining accounts and the expertise of scientists and professionals who write of their infinite explorations and curious finds. It is made with bold, colorful photographs of places that most people will see only in images—Greenland's ice caves, the Yucatán's liquid blue cenotes, the Grand Canyon's unreachable pinnacles, planes, and hidden passages.

Here are places far beyond the wilderness Lewis and Clark experienced, where one can sense the true essence of wilderness, where the human voice carries over pools of water so old they never saw the reflections of humans until the speaker came to be there. Here also are such people as the legendary western caver Donald Davis, who, in addition to his comical cave dilation theory, speaks of cavers as notes in a symphony being played against a grand backdrop of timelessness. Like many cavers, he is the kind of person who inspires explorers to greatness.

In these pages you will feel the power of caves on your imagination; you will begin to understand what I mean when I say that I have been in sacred places, none of them constructed by people, that I have heard whispers of things gone by while sitting alone in the old hollow places. I have heard the drumming of my heart while surrounded by solid stone, shirt off and skin wet with sweat, and felt the coolness of rock hard against my chest and back, the heat rising in my face as I struggle to move. I have felt the lure of stone passages, filled with water to the top, seen the beckoning of the blue, heard whispered the promise to let me go, if only I will risk a little more. I have felt the pull of the darkness from the lip of a great open pit beneath the earth, and wondered just for a moment why there are people who will never take the first step—from which all great journeys begin.

I have seen the constructions of people and know them to be simple, and not very humble, attempts to create holy places, and I have not been fooled.

Read on.

RONAL C. KERBO
NSS 11539
Colorado

Ronal C. Kerbo is the senior cave specialist of the United States National Park Service.

WHY CAVING?

Just north of the Arctic Circle, at approximately the same longitude as Rio de Janeiro, Hazel Barton and Nancy Holler Aulenbach stood atop an ice cliff, staring into a deep blue ravine that was spanned by a nylon strand only nine millimeters thick. The line swayed like an incredibly long jump rope as a team member clipped it to a metal screw drilled eight inches into Greenland's polar ice cap. Nearly 150 feet below them, at the base of the fissure, a river boiled within its frigid channel. The rare above-freezing day had contributed fresh runoff to the unnamed torrent. A short distance away the river vanished into the Malik Moulin, a yawning 700-foot-deep chasm in the ice. This frozen cave was their immediate goal.

To reach the glacial pit and the deep passage that lay beyond it, the two would have to commit to sliding down the slender rope into the empty space beneath their dangling feet. Perhaps the most nerve-racking aspect of the procedure was that they would have to smile cheerfully into the eye of the IMAX camera as they took the heart-pounding step away from the cliff.

Neither woman was a stranger to potentially dangerous rope work. Nancy and Hazel, both 27, were expert cave explorers— or cavers, as they are typically called. Both

had faced deeper descents in their long caving careers. But they were used to climbing and rappelling over limestone deep within the belly of the Earth, where a pit was nothing more than a blackness that swallowed the beam of your cap lamp when you looked down. Here on the ice cliff, the watery polar sun sparkled off a slick, treacherous surface and reached far into the depths of the ice, painting everything below a brilliant blue. If you fell, you would see the bottom coming at you from a long way off.

Hazel opened a small nylon pack and began removing her climbing harness and a French rappel stop. The spring-loaded device was designed to control the speed of a long rope decent. A few feet away, Nancy pulled out an American rappel rack—a U-shaped metal bar designed to perform the same task. Puffs of dust rose from her well-worn gear as it hit the ice, traces of mud from an Alabama cave she had visited the week before. It seemed hard to believe that she had met Hazel on an international flight just 24 hours earlier.

The two explorers had flown from Washington, D.C. to Copenhagen, only to catch a flight back across the North Atlantic to Kangerlussuaq, Greenland. There they were quickly hustled onto a Sikorsky helicopter and flown to the permanent ice cap,

where they met the crew of French ice cavers and American filmmakers whose expedition they had joined. When welcoming Hazel and Nancy to their makeshift camp atop the glacier, team leader Janot Lamberton warned them of the many hazards of entering ice caves in a French litany that neither of them understood. Another team member translated: Ice explodes—frozen chunks the size of major appliances can shoot out of cave walls without warning, propelled by the immense pressure of the overlying ice cap. As if to punctuate his remarks, popping ice, like a series of shotgun blasts, could be heard in the distance.

The cavers spent an uncomfortable night on the ice, 50-knot winds howling beneath a clear sky lit by the glowing green bands of the aurora borealis. When Hazel first braved a walk through the wind to the camp's latrine (a toilet seat wedged above a small fissure), a nearby ice explosion showered her with shards. The next morning, jet-lagged and exhausted, the two women fastened their harnesses with numb fingers, all the while still struggling with a language barrier that was making it difficult to understand who Janot thought should be the first to rig into the traverse. If all went as planned, though, within an hour they would be exploring the blue world below, ducking bursting walls as they surveyed the ever changing drainage system that carried water beneath the ice.

To a casual observer—if a casual observer could somehow stand there, atop a slick ledge far above a void swallowing an enormous river, surrounded by thousands of miles of jagged glacier—what Nancy and Hazel were about to do might have appeared dangerously foolhardy. Such was the opinion that many held of Erik the Red who, according to ancient Icelandic texts, in 985 ventured toward unknown, iceberg-filled seas to the west. Despite dire predictions, he found the world's largest island: 1,600 miles long by 700 miles wide. Deep fjords and

Led by the yellow glow of a carbide lamp, caver Jim Hewitt negotiates a typical muddy squeeze. This one is named MHTE—Most Horrible Thing Ever.

mossy coves rich in sea life rimmed a seemingly endless plain of undulating ice. Erik named the place Greenland, and established a Norse presence there that remained for centuries, evidenced today by crumbling stone churches and lodges. While civilization tenuously clung to its fringes, Greenland's vast interior, crushed under a mile-thick ice sheet, remained a mystery.

With both the caution and confidence of Erik launching his longboat into a frigid sea, Hazel was the first to approach the swaying rope.

No matter how forbidding, how plainly dangerous the geography of a given place, there is something inherent in human nature that will push us to explore. The earliest records of our species were painted by flickering torchlight on cave walls, far from sunlight and safety. Long before history began, humanity moved out of Africa and across the globe, driven by a hunger to know what lay in the territory beyond.

The hunger is seldom appeased today.

Travel to a remote village of the Amazon or equatorial Africa and you are likely to be greeted by people wearing Nikes® and Pokémon® T-shirts. Walk into any department store, and you can purchase a handheld global positioning device that will tell you exactly where you stand. Go on-line, and you can instantly access detailed satellite photographs of any real estate on the surface of the planet, along with sonar soundings of the deepest ocean trenches. The ancient Earth remains infinitely varied in its terrains and habitats, offering more landscapes than the most diligent traveler could sample in a lifetime. But for the true explorer—the carrier of that uniquely human gene that drives her or him to proclaim, I must be the first one there—the ever shrinking planet offers only one remaining option: look underground, where not even the satellites can reach.

The motivation to explore caves is different from perhaps that, say, to climb Mount Everest. It's not an urge to conquer the cave because it's there; it exists. Caving is driven by a curiosity and desire to know, precisely, what lies beyond each yawning entrance and forking passage. Caves have no "there" to them until they are explored. In that sense, even a flashlight-toting weekend spelunker shares the ancient human need to reveal new ground.

Millions of caves hide beneath the surface of the world, and over the past half century, thousands of expert cavers, armed with modern technology, have begun to map and explore a small percentage of them. Natural tunnels and chambers can be found in limestone and other sedimentary rocks, in volcanoes and lava tubes, in glaciers and polar ice. With ropes and highly specialized climbing gear, cavers routinely visit pits over a thousand feet deep and haul underground base camps through bone-hugging crawlways thousands of feet long. Cave divers, using technology so sophisticated that they might as well be spacewalkers, now swim through miles of flooded passages, pulled along by lamplit electric scooters through underground rooms born in darkness, which until that day had never been exposed to light. Unlike explorers of ages past who sought gold or conquest or new lands to exploit, modern cavers display a fervent respect for the places they discover. Caves and the life within them are often extraordinarily delicate, easily disturbed by human intrusion. Cavers thus take great pains to make the evidence of their passage all but invisible. And yet, like the European explorers of the 16th century, cavers have often traversed what they considered empty wilderness, unaware they were harming indigenous populations. Cave biologists have long studied and sought to protect bats, fish, insects, and other familiar cave life, but it is now known that there are far more extensive—and unfamiliar—populations of life that exist on and within the Earth that have yet to be identified.

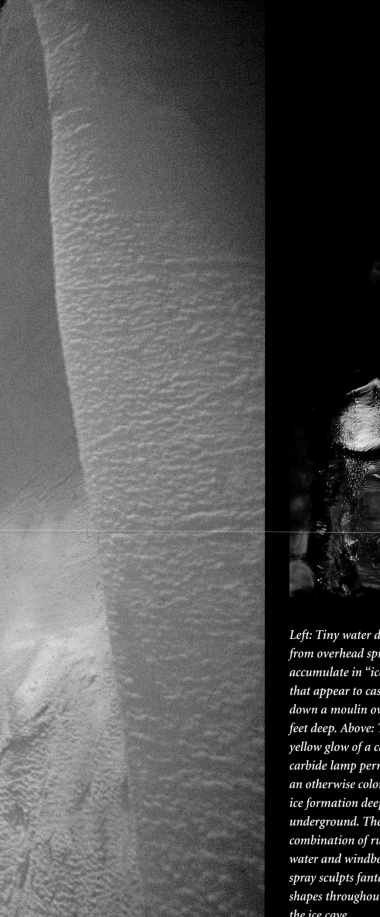

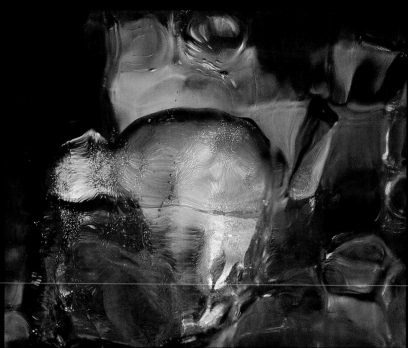

Left: Tiny water droplets from overhead spray accumulate in "ice balls" that appear to cascade down a moulin over 300 feet deep. Above: The yellow glow of a caver's carbide lamp permeates an otherwise colorless ice formation deep underground. The combination of rushing water and windborne spray sculpts fantastic shapes throughout the ice cave.

Kyerksgöll, an unusual glacial cave in Iceland, was created by heat rising from a volcano. As the glacier slides over the underlying heat source, the cave slowly grows. The volcano provides sulfur and other minerals that give the ice an orange tint—and could serve as energy sources for unseen microbial ecosystems.

CHAPTER I

A VAST TWILIT WORLD

The skin of the world hides many caves. All are profoundly shaped by, and profoundly affect, the nature of the land overhead. Caves are to a surface landscape as veins and capillaries are to a human face—the hidden structure of an inseparable whole. Whether the more familiar natural tunnels of eroded limestone, the lava tubes of volcanic landscapes, or the ice caves that twist beneath living glaciers, the surface gives explorers clues to the shape and size of what lies below. Finding a significant new cave is an act of geographical deduction: it requires the cave hunter to know the lay of the land. ■ One of the first Europeans to systematically study the geography of Greenland's ice was a teenager from Denmark named Peter Freuchen. At 18 he was a bright, competent medical student at the University of Copenhagen. Freuchen had grown up in a coastal village, his muscles hardened and hands callused from occasional shipboard jobs. Tall, broad-shouldered, and well mannered, he was popular with his professors and their daughters. There was a brooding, distant quality to the young man's face befitting his romantic attachment to the sea. In his spare time he could be found listening to sailors' tales at the docks or sailing alone in the small boat he had owned since he was eight, eyes scanning the horizon. ■ Freuchen was fascinated by the natural world and briefly entertained the idea of becoming a geologist or biologist. But he had decided, as he recounted 30 years later in his

Above: Peter Freuchen in 1921, after living on the ice for over two decades. Right: Packed snow forms a loose ceiling in many Greenland ice caves. In warmer weather, this can collapse without warning.

An unidentified member of Peter Freuchen's 1909 expedition stands before the face of the Greenland ice cap. The dark bands correspond to past summers, when algae would briefly bloom atop the ice. By measuring the thickness of individual bands, modern scientists can plot past climate changes.

classic memoir *Arctic Adventure,* that "such research is for the dilettante or the young man of means, and I knew I would have to make my own way."

Medical science was making great strides in 1905, and Freuchen felt that a future as a doctor would be secure. Undoubtedly it would have been, if not for an emergency-room experience that shook the student's faith in medicine and ultimately propelled him away from the technological century growing up around him and into the frozen, ancient world that would become his home. A dockhand was crushed in an accident while offloading a freighter. Skull fractured beyond recognition, limbs and rib cage all

but destroyed, the man was at first taken for dead. But a bystander detected a feeble heartbeat, and young Freuchen, on duty at the Copenhagen hospital, assisted several doctors in a lengthy surgery.

All thought the case hopeless. The man had lost too much blood, had suffered too many injuries to survive. And yet he did. For six months Freuchen watched the patient's slow recovery, amazed at the human will to survive and the body's power to heal. On the day the sailor was released, surgeons from around Europe gathered in Copenhagen to congratulate the staff and patient. All stood solemnly watching as the man slowly made his way from his room to the door of the hospital. He walked down the sidewalk and into the street where—in what could have been the punch line of some poor cosmic joke—he was promptly run over and killed by one of Copenhagen's first automobiles.

The incident "made me burn with impotent fury," Freuchen wrote. "I decided I was not cut out to be a doctor, and I left school immediately."

By chance the Danish government was sponsoring an expedition to explore and map northern Greenland, and Freuchen pestered the leader until he was accepted as a member of the team. Like Eric the Red exiled from Norway a millennium earlier, Freuchen embraced the North Country. First he volunteered to travel alone ahead of the expedition to purchase dogs from the native Inuit, toward whom he felt an inexplicable attraction. Despite the daily hardships of Inuit life, the eyes of the people Freuchen befriended always seemed alive with humor behind their otherwise impassive faces. Winter winds froze their skin and the summer sun burned it until their faces glowed like patent leather. Men and women alike seemed to judge a stranger not by who he was or what he owned, but by his willingness to learn and to work toward a common good. Freuchen saw freedom and democracy in such attitudes. For the rest of his life, he would study and follow Inuit ways.

A year after arriving in Greenland, he volunteered to staff the first research station to be built at the edge of the permanent ice cap. When the small wooden cabin was completed, it measured about 9-by-15 feet. But almost immediately it began to shrink as layers of ice from the researchers' breath condensed on the walls.

"Toward the last the room became so small that two men could not pass each other without rubbing elbows," Freuchen recalled.

At first accompanied by two other volunteers, he spent close to a year studying the ice cap and its relationship to the underlying geology. During the long Arctic winter,

his housemates fled with a supply ship. Freuchen stayed on alone. For six months he was surrounded by wind that howled unchecked over the ice. Nightly he fought off roving packs of wolves that ultimately consumed his dogs and many of his stores.

At noon every day in October, a few slanted rays of the vanishing sun lit a nearby peak. Freuchen would climb to the top each day in order to survey his vast twilit world, recording the movement of the ice at three levels in the 3000-foot face of the cap. He came to know each foothold, crevasse, and cave intimately. When the total darkness of

In 1912, Freuchen joined Knud Rasmussen in the first expedition to set out on foot from "Ultima Thule," the most remote northern land, eastward across the ice cap. (North is to the left in this map.) Freuchen would ultimately make five such journeys before losing a limb to frostbite.

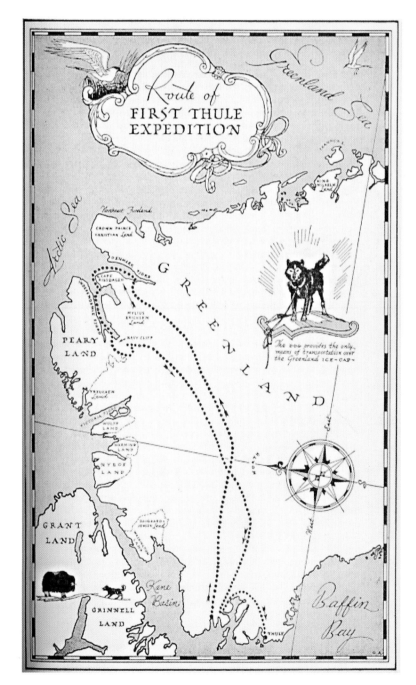

winter fell like death itself, he could make his way to the top solely by feel.

Greenland's ice cap is thickest in the northeast quadrant of the tusk-shaped island. The precipitation is greatest about 100 miles due west of Pustervig, the eastern bay where Freuchen manned his research station. There the domed cap approaches a depth of two miles. Some 300 trillion tons of the ice press the center of Greenland down into the Earth's crust—remove the cap and the land would literally bounce upward.

The term "solid ice" is a misnomer. Like aged glass, a thick mass of ice will always flow downhill, at a pace too slow to register on a human time frame. The Greenland cap pours outward in all directions from its central high point, rising and falling with the underlying bedrock. Where the base contains ridges and mountains, the ice forms peaks and crags above, a jumble of gigantic blocks sculpted into fantastic shapes by the relentless wind. Where the ice flows over a smooth plain, it forms a gentle pattern of parallel rolling hills, like the sand ridges beneath the waves of a gently sloped beach. In the valleys between such ridges, summer meltwater gathers, forming rivers that run straight for miles, scouring canyons in the ice. During cold summer nights, the surface of these rivers freeze over while the water below continues to move seaward. In winter, the rivers freeze solid in their carved beds. Thus over time in these ice valleys, long horizontal caves are formed, roofed by the packed snows of winter, their walls lined by the banded layers of ice from within the ancient cap itself.

Occasionally such rivers will encounter a fault in the ice, usually corresponding to change in the underlying geology. Then the rivers will plunge downward, forming pits or moulins (a French term meaning "water well") hundreds of feet deep. The hidden rivers emerge at the perimeter of the cap, each forming the headwaters of one of Greenland's many fjords. This is why steep, fertile valleys, rich in wildlife and geologically stable, are so common around the edges of the island, while the interior remains an unknown, ever changing body of living ice. The ice cap depicts geology in fast-forward mode. Hydrologic processes that take several million years to evolve on land can happen there in a matter of days or weeks.

As winter comes and the rivers recede, the upper entrances to the uncharted cave systems may freeze shut. The ice itself continues to flow at its sedate pace of a few inches per year. Yet over the course of a season, this movement is enough to cause many of the new caves to collapse. Thus each spring the process begins anew, with the rivers sometimes entering old caves, other times excavating new ones. The changeable nature of this environment makes it difficult—and dangerous—to study close up. Cavers scoff at typical Hollywood depictions of cave-ins and collapses in natural caverns because caves in limestone tend to be geologically stable, changing little over centuries. But the ice caves of Greenland are if anything shakier than those in movies; unlike terrestrial caves, a bad step or a loud shout inside the ice really can bring the walls tumbling down. Despite the dangers, such caves are worthy of study because, among other treasures, a hidden history of the Earth's climate is written in the walls.

A 1910 photograph shows Freuchen standing at the base of an exposed cliff of ancient ice in one of the river canyons. Black and white horizontal bands mark the scalloped sides. Freuchen recognized that the dark lines corresponded to summer, when the surface of the ice constantly melts and refreezes, creating shallow, algae-darkened pools that dotted the landscape for hundreds of miles. The white lines corresponded to the snows of winter. By counting the number of black and white bands as one moved down into the ice cap, Freuchen realized that one could read the climates of past decades and centuries.

Scientists use the technique today.

A study of bands in the Greenland ice that was published in 1998 in the journal *Science* revealed temperature fluctuations of the past several thousand years. According to the study, 5,000 years ago, when the Inuit first flourished in the region, temperatures were 4.5° F warmer than present. When Eric the Red first established settlements in Greenland a thousand years ago, it was 2° warmer than today. In contrast, during the medieval little ice age—corresponding to the time when Viking settlements mysteriously vanished—temperatures averaged about 2° colder than now. And in the last true ice age, some 22,000 years ago, average temperatures plummeted to a whopping 41° below current levels.

The ice cap may give clues as to future climactic change, as well. A 1999 study by NASA scientists confirmed that Greenland's ice sheet had grown thinner over the previous decade. Glaciers along the southeastern coast are thinning, perhaps due to global warming. Researchers comparing aerial surveys of the ice sheet taken in 1993 and 1994 with a similar survey taken in 1998 concluded that parts of the sheet near the ocean diminished at a rate of more than three feet per year. While ice layers thickened by up to ten inches in interior regions during the same period, the net effect was a loss of mass for the entire cap.

The study found that glaciers were also moving into the ocean at a higher rate, which in the future could significantly raise the

A solitary figure moves toward base camp at sunset. Team members carried radios and global positioning system units to avoid becoming lost amid the sameness of the undulating plains of ice.

Greenland's ice caves eventually empty into long, narrow fjords that form fertile coastal valleys rich in plant life. Fresh water and minerals carried from glaciers fertilize the rocky soil, creating pasture for reindeer and other wildlife.

oceans during a period of global warming.

Freuchen depended on Greenland's ice caves for warming of a different sort.

On Christmas Day 1910, he and a group of Inuit hunters were trapped on thin ice while attempting to dogsled to the island of Tasiusaq along the western shore. Separated from the others on a floating ice pan, Freuchen drifted alone, convinced he would die. Occasionally he would hear a distant shout from one of his companions, trapped on icy rafts of their own, but for hours he could neither see nor communicate with another human. Meanwhile, the temperature dropped far below zero.

Later, an Inuit shaman named Asayuk drifted close enough that Freuchen was able to join him. Throughout the long night, the old man waited for subtle signs of any ice that might safely hold their weight. Just before dawn he decided they were as close to shore as they would get and the ice was as strong as they could hope to find. The two jumped and scrambled for shore, only to crash through. Luckily, the tide had receded; the chilled seawater was only a few feet deep. The soaked men soon heard shouts from other members of their party. They all made their way to an ice cave, where the rest had already camped and built a small fire.

At one point as the shivering men recovered, an immense ice block burst from the wall and crashed near Freuchen. He leaped up, thinking that the fire had begun to melt the walls, and prepared to run back into the cold. Asayuk explained that the spirit of the mountain demanded darkness for his slumber because it was winter. The spirit would not harm anyone if the interlopers left him some meat when they departed. Freuchen protested, but Asayuk interpreted the fact that none of them had actually been hit as proof that the spirit was friendly.

"I tried to explain the scientific reason for the blast," Freuchen recalled. "But Asayuk looked at me as a man does at a child who is unable to grasp eternal verities. He explained patiently that he had heard the same expla-

level of the world's oceans, according to Bill Krabill, a researcher at NASA's Goddard Space Flight Center. Krabill guessed that the glaciers might be speeding their flow to the ocean because more melted ice on the surface is seeping to the bedrock, and the massive ice mountains are sliding upon this liquid layer. Understanding the ice caves that transport the water could thus provide insight into how to gauge the rise of the nations from Verhoeff, a stone-wise gentleman who had followed Peary. And Verhoeff had disappeared and never been found—the spirit of the mountains had taken him to show him what was the truth and what was not the truth."

Freuchen continued to travel Greenland and, along with his friend, the famed explorer Knud Rasmussen, to learn the ways of its native people. He married an Inuit woman named Navarana—the subject of several of Freuchen's later books—and with her fathered two children. She died in an influenza epidemic in 1921. Then, in 1923, while on expedition to the west coast of Baffin Island, Freuchen faced a spirit that was far less benevolent than the one he'd encountered in the ice cave 13 years earlier.

Separated from his party in a blinding blizzard, Freuchen dug a shallow trench in the snow and pulled his sled over the top. Exhausted, he collapsed into sleep. On waking, he had no feeling in his left foot. When he tried to move, he found his sled was frozen in place above him. Eventually he moistened a length of bearskin by chewing on it; it froze hard as iron. Using this improvised tool, he managed to scrape a small opening in the snow and pull himself from beneath the sled.

A hunting party saved Freuchen, but gangrene set in when his foot thawed. The flesh around his toes fell away until the bones protruded. The shaman treating him wanted to remove the toes with her teeth to prevent dark spirits from entering his body. Freuchen instead chose to knock them off himself with a hammer. He eventually lost the foot and was given a peg leg of elaborately carved walrus tusk. After returning to Denmark with his son, Freuchen began a new life as a writer, chronicling his time in Greenland in several nonfiction books, novels, and films.

Having lived in peace and freedom with native peoples, from whom he had learned much, he became fiercely anti-Nazi and worked for the Danish underground in Copenhagen during World War II. The German occupation government captured Freuchen and sentenced him to death, but he managed to escape to the United States, where he lived peacefully until 1957. His son returned to Greenland and his great-grandchildren live there still.

At about the same time that Freuchen fled the Nazis, the United States Air Force commissioned Søndrestrømfjord, or Sondrestrom Air Base, code name Bluie West 8. It was sited at the head of one of Greenland's longest fjords, where it was less likely to be affected by coastal storms. After the German occupation of Denmark, Greenland's security had been entrusted to the U.S. by the Danish Ambassador in Washington, Henrik Kauffmann. Bluie West 8 soon became one of the most important refueling sites for missions flying between America and its allies in Europe, owing to the generally good weather for which the airstrip became known.

After the war, the base became important for servicing the radar systems designed to provide early warning against a Soviet nuclear missile attack. Between 1954 and

Musk oxen huddle together in North Peary Land in Northeast Greenland National Park. The large animals share the park with reindeer, arctic foxes, alpine hares, and a surprisingly large bird population.

1965, the airline SAS also began making use of Søndrestrømfjord for stopovers on the long route between Copenhagen and Los Angeles. With a civilian link thus created, the airport eventually became a travel gateway to Greenland. When the Cold War began to collapse in 1989, the Pentagon decided to shut down its Greenland radar stations and close the American base. In 1992, the airport came under Greenland home rule and was given its first Greenlandic name, Kangerlussuaq—meaning "long fjord."

The fjord is over 100 miles long, with Kangerlussuaq Airport located at its far end, just a few miles from the edge of the permanent ice cap. Some 90 percent of all travelers to Greenland now pass through Kangerlussuaq, a town still resembling the remote military outpost from which it grew. With a population of just 325, Kangerlussuaq Airport remains exclusively a civil aviation area, outside any municipal classification. The resident population is directly or indirectly employed by airport operations.

Just beyond the clustered barracklike metal buildings, one can find wild reindeer, musk oxen, arctic foxes, alpine hare, grouse, gyrfalcons, and peregrines. The fjord is home to Greenland cod and a type of trout called char. The countryside around Kangerlussuaq turns green during the short arctic summer, with a surprisingly rich flora. But from November to mid-June, the long fjord freezes over. The sun ducks below the horizon, and does not return for two months.

Although the days were shortening in late September 1998, sunlight still sparkled off the distant edge of the ice cap when cavers Hazel Barton and Nancy Aulenbach stepped from the airliner, still woozy from their long flight—not to mention first-class bar service—to find themselves hustled into heavy Gore-Tex® suits and onto a noisy Sikorsky that waited to ferry them to where they could meet the spirits of the mountain up close and in person.

Like most Inuit villages in Greenland, Tunu, near Angmagssalik, perches upon a rocky fjord at the edge of the ice cap. The island is ringed by long valleys that have accommodated virtually all of its population since before the arrival of Vikings in the tenth century. The immense interior remains as forbidding and empty now as it was to Erik the Red.

CHAPTER 2

BREAKING THE ICE

A scowl on his face, Janot Lamberton approached Nancy Aulenbach. "Non!" he shouted, pointing at her rappel rack, spouting invective in rapid-fire French Nancy couldn't comprehend. The tone of the message, however, was as clear as the explorer's angry blue eyes: he did not approve of her rig. Janot called over his son Mael, a strong caver in his own right, to translate. The expedition stood on the ice, listening to the roar of the blue river in its channel below as Mael approached. The film crew shifted nervously. Hazel paused in pulling on her seat harness. ■ "This is not safe," Mael said after listening to his father. "You must wear one like these." He held up the Petzl-brand rappel bobbin, also called a "stop," that was clipped to his own harness. ■ "No way," Nancy said. "I've been doing pits my whole life with a rappel rack, and I'm not going to change now." ■ "Then you must get off of our rope." ■ They glared at each other, two highly competent experts, neither willing to back down. ■ Technically, neither the Petzl nor the rack was required for the traverse, since the object was not to rappel down a slack rope but to move horizontally across a taut line. But in every maneuver on rope, an experienced caver must be prepared to ascend or descend, should an unexpected need to do so arise. It is one of the first things you learned when studying SRT.

Above: Nancy, right, shares a light moment in the mess tent with Hazel. Right: A caver rappels into an undercut moulin. Cavers often had to climb out of the pits by midday, before the spray from afternoon snow melt encased ropes in ice.

Runoff from dozens of freshets gathers into rivers that may flow miles over the ice cap before plunging into moulins such as this one. Winter snows can sometimes bury the frozen rivers, which begin to flow once again in spring, creating long horizontal caves that end in sudden pits.

Short for single rope technique, SRT is the method of choice for descending vertical caves. Replacing older systems involving cable ladders or winches, SRT evolved differently in Europe and the United States in the 1960s and 70s, resulting in slightly different types of gear designed to perform identical tasks. While many American cavers have adopted European rigs and climbing methods for international expeditions, those from the premier U.S. caving area known as TAG—an acronym for Tennessee, Alabama, and Georgia—remain loyal to the SRT style first used to conquer its deep pits. Nancy lived in Atlanta; her parents, both avid cavers, had lowered her down her first TAG pit when she was in diapers. Like Mael, she had grown up thinking of family outings as a chance to get on rope underground. She wasn't about switch to an unfamiliar device in order to please a group of angry Frenchmen.

"Fine," she said, unclipping her Jumar ascender from the rope, loudly rattling the aluminum break bars of her rack. "Do the traverse without me." She walked away from the cliff, her crampons grating on the ice.

Hazel, a Brit who began caving with a youth club in Bristol at age 14, did have a European bobbin. Janot suggested that Hazel and Mael traverse above the river in the daylight that remained; he and Nancy would address the gear debate later. Janot had brought his son to Greenland on his first expedition in 1985, when Mael was 16. The boy had fallen in love with the place. Like Peter Freuchen before him, Mael had eventually fallen in love with an Inuit woman as well. They now had two sons with traditional Inuit names: Minnik and Malik.

Photographer Gordon Brown prepared to run a thick reel of large-format film—representing only about 90 seconds of footage—through the same IMAX Mark II camera that had ascended Mount Everest two years before. Crew member Dave Shultz chopped an additional anchor for the camera tripod with a large ice ax. A few years earlier, Shultz had helped the actor Sylvester Stallone make a similar traverse for the film *Cliffhanger*. It was Stallone's ice ax from the film that he now used.

"Okay, we're loaded," Brown said

With that, Hazel stepped from the ice and began sliding along the rope, which swayed slightly above a turbulent froth more than a hundred feet below. The water was the breathtakingly clear blue one would expect of a Caribbean lagoon or the pool of a fashionable hotel. It roiled within its channel, suggesting to Hazel a delightful scene for a kayak run—until, that is, the unfortunate kayaker encountered the moulin.

Luc had explained to her that a deeper channel meant the river had followed the same course for a greater number of seasons. This channel, approaching 200 feet deep in places, was unusually old for the ice, perhaps 15 years or more. Such a large volume of water dropping into the same pit for 15 summers meant the cave somewhere below would have to be incredibly large. Its lowest reaches would meander through ice tens of thousands of years old. As Hazel inched along the rope, she wondered whether the weather would allow her to search for frozen microorganisms in the ancient ice, her working goal on this expedition.

On this first day of the expedition, tensions among the group were nearly as taut as the line from which they dangled. The dispute over Nancy's climbing system was part of a pattern, a symptom of the team's general unease over conditions that they could not control. The huge volume of water sluicing down the channel and into the nearby moulin was far more than anyone had expected. It was too wet for anyone to safely enter the cave. Today they might travel a short distance into a horizontal entrance above the river, but the expedition's major objectives were, for now, off limits.

Since he first began leading expeditions to Greenland in 1985, Janot Lamberton, the undisputed master of deep ice caves, had never seen weather so warm in late September. A few days before the Americans had arrived, Janot had descended part of the way into Malik, a pit over 600 feet deep named after his grandson, to find it nearly dry. Now water thundered down the shaft in a pounding spray that could kill a caver on rope in a matter of minutes. If the sheer force of the water didn't rip you off the rope, the hypothermia would numb your brain until you stopped climbing and simply fell asleep in the deadly spray. Even breathing in such conditions becomes hazardous, because the spray throws off tiny ice crystals that could damage the lungs, eventually suffocating a climber. Both the Americans and the French had hoped to set a new world record for deep descent of

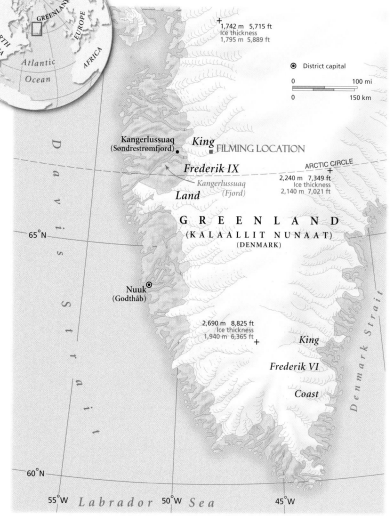

an ice cave some time during the next two weeks, but for even a possibility of that, the weather would have to turn much colder.

A similarly frustrating late-season warming had occurred during Janot's 1996 expedition, turning his camp into a shallow lake and keeping his scientific team out of the larger caves. Luc Moreau, a well-known glaciologist, had been on that trip. He had joined the current team in hopes of finding

a way to the depths he had missed two years before.

"There are very few places on Earth where we can actually enter a glacier and study the interior," Luc had told Nancy in camp. "We know quite a bit about the surface of glaciers, but very little about what happens inside."

What was happening amid the powerful river currents that coursed through the cave could provide clues to the way global warming might affect the ice cap, and thus ultimately the world's oceans. The 1996 expedition had taken place in August; even then, water levels had seemed unusually high for late summer. Now, in late September, both Luc and Janot had fully expected surface streams to be frozen solid, and the cave watercourses to have faded to easily avoided trickles.

The surging force below Hazel's feet was anything but a trickle.

Earlier that day, she and Nancy had helped avert a disaster caused by the warming ice. Kim, an expatriate Dane living in Greenland, had brought his snowmobile and trailer from Kangerlussuaq in order to help ferry filming and scientific equipment to and from camp and to various locations on the ice. After the long trip and the welcoming party the night before, Hazel and Nancy were adjusting to what would become a daily routine. In the subfreezing morning temperatures, they donned several layers of expedition-weight fleece, bright florescent Gore-Tex® oversuits, glove liners, gloves, boots, and crampons. Neither had worn crampons before, but they quickly learned it was impossible to stand on the slick ice without them. That morning the two were the last to walk toward the canyon, where most of the rest of the expedition was already scouting and rigging the traverse.

As they hiked through ice and the shallow, algae-rimmed puddles already beginning to form in the rising sun, they heard the snowmobile struggling nearby, its engine revving wildly. Hazel ran over a short

34

rise and saw that the trailer, loaded down with thousands of dollars worth of gear, had broken the ice, which collapsed into several feet of water—and the trailer was slowly pulling Kim and the snowmobile backward along a tilted slab.

Nancy plunged into the waist-deep water behind the trailer; Hazel jumped in after her. Together they shoved the back of the trailer, their feet sliding under water along the slab. Kim throttled the snowmobile forward. With the cavers digging their crampons into the submerged ice, the machine leaped onto solid ground, saving the vehicle and gear.

It may be easier to understand a drive for extreme exploration in those born to it, like Nancy Aulenbach and Mael Lamberton, than in those who willfully seek it out. Leif Eriksson might never have discovered Vinland, the continent eventually called North America, were his father not Erik the Red. But what pushes first-generation explorers to the edge? Why do they give up the familiar, the comfortable, in favor of unknown danger? Are they escaping past events or merely boredom? According to ancient texts, Erik the Red had been exiled because of "some murders" in Norway, then ran into disputes with his neighbors in Iceland. Peter Freuchen always claimed that the sudden death of his patient had propelled him from a comfortable life as a Danish physician. For me, the lure of the cave was a childhood fantasy, dug from holes in the sandy Florida soil of my backyard. But what first drove a Bristol barmaid named Hazel Barton underground?

She was born in August 1971 to the manager of a grocery in Bristol and to a saleswoman in a yarn shop. Hazel describes her mother as "a proper English mum, always crocheting something for the church auxiliary sale." The middle of three children, Hazel was curious about everything. At the age of four she conducted a scientific

Continued on page 40

The expedition's tents stand as the highest objects for hundreds of miles. Arctic winds, whipping over the endless plains of ice, assaulted the tents nightly. Incredibly, only one collapsed during the course of the expedition.

A caver peers down a "window" into Moulin Afrecaene during Lamberton's 1996 expedition. The caves can change each night as fresh runoff or the spray from a waterfall freezes. Sometimes cavers find ropes they used the day before encased in solid ice; other times what had been an open shaft is covered by an eerily thin cap of ice, blackness still visible beneath it. By the 1998 expedition, this moulin had vanished.

BUGS FROM SPACE?

BY MICHAEL RAY TAYLOR

In the rough-and-tumble world of bacteria, *Streptococcus mitis* qualifies as a wimp. The ball-shaped microbe prefers the easy life, hanging out in warm, moist environments loaded with sweets. Your mouth is ideal—there's probably a few hundred million of the species in there right now, eagerly awaiting your next meal. But swish a bit of mouthwash and wimpy *S. mitis* cells rupture and die by the score.

So it came as something of a shock to NASA scientists when a colony of *S. mitis*, inadvertently sent to the moon aboard the *Surveyor 3* probe in 1967, was brought back alive by Apollo 12 astronauts two-and-a-half years later. In late 1969, microbiologists cultured nearly a hundred live cells from a cubic centimeter of a foam camera housing that astronaut Pete Conrad had retrieved from the *Surveyor 3* probe under sterile conditions. The moon offers one of the most hostile environments imaginable. The vacuum alone would kill a human in seconds. Despite an average temperature of only 20° above absolute zero, the lunar surface is constantly bombarded with solar radiation that can scramble DNA into useless acid.

As microbial habitats go, it's a long way from your mouth.

Yet in 1969, scientists considered it little more than a curiosity that the microbes, which must have been deposited in the probe during assembly, had somehow survived their long journey. Upon reviewing transcripts of his lunar conversations in 1990, Conrad lamented, "I always thought the most significant thing we found on the whole Moon was that little

These bacteria feed an animal that cannot eat, a tube worm of a deep Pacific volcanic vent.

bacteria who came back and lived."

At last scientists are starting to talk. They're discussing not only the little "bug" that hitchhiked to the moon, but a host of hardier Earth bacteria called extremophiles—microbes that "love" extreme conditions of heat, acidity, salinity, pressure, and other environmental constraints deadly to more familiar life. Collectively these bugs are providing evidence that microbes might survive the harsh conditions of other worlds in our solar system or even long periods of travel in deep space.

The past three decades have seen a paradigm-shifting revolution in microbiology, as places once thought incapable of sustaining life have been shown to teem with it. In 1977, researchers working in the research submarine *Alvin* deep in the Pacific discovered unexpected tons of microbes spewing up from the volcanic vents of the sea floor, fed by a hot chemical soup. Since then, extremophiles have been cultured from Antarctic ice; deep caves, mines, and oil wells; nuclear-reactor cores; and subterranean pools containing supposedly "deadly" concentrations of acids, salts, and alkalis.

When conditions become too dry or cold, some bacteria enter a state of suspended animation, surviving for incredibly long times. In a recent article on pychrophiles—cold-loving microbes—in the *Proceedings of the National Academy of Science*, P. Buford Price of University of California, Berkeley, describes viable specimens of a spore-forming *bacillus* and of an extremely halophilic (or salt-loving) bacterium revived from inside a salt crystal 250 million years old. In another case a bacterial spore was revived, cultured, and identified from a 40-million-year-old hunk of amber. The amazing abilities of such bugs are bringing respect to what was once regarded as a fringe theory: panspermia, or the seeding of microbial life from one planet to another.

"We now know of several terrestrial microbes that can pretty well survive any of the so-called 'sterilizing' elements of space travel," says Richard Hoover, a chemist and microbiologist with NASA's Marshall Space Flight Center. Hoover and other NASA scientists have recently been extracting bugs from deep, ancient ice in Antarctica, Siberia, Alaska, and Greenland to study their long-term survival tactics.

Meaning literally "seeds everywhere," panspermia's earliest recorded advocate was the ancient Greek philosopher Anaxagoras. In the 1970s, British astronomers Fred Hoyle and Chandra Wickramasinghe revived the theory, arguing that concentrations of organic chemicals in interstellar dust were evidence that bacterial spores floated about the cosmos. The notion provided fodder for bad science-fiction movies, and for years nearly all biologists considered it laughable. Not only did it seem impossible for bugs to survive unprotected space travel, but few scientists could envision a realistic scenario for getting microbes off a planet's surface.

As microbiologists were beginning to understand life's extremes, planetary scientists were getting a handle on the mechanics of large impacts, collisions of the sort that famously killed off Earth's dinosaurs 65 million years ago. Study of impact craters on the Earth, the moon, and other planetary bodies, along with increasingly sophisticated computer models, suggest that on the edges of every big collision, anywhere from a few pounds to a few tons of dirt and rock are launched into orbit. If the launched rocks happen to contain the right extremophiles, they could become natural spacecraft, never heated, stressed, chilled, or radiated to the point of killing their small passengers as they drift in space for

Astrobiologist Richard Hoover gave the name "The Klingon" to this unidentified microbe.

millions of years.

That leaves the question of whether bugs in such capsules could survive entry through a planet's atmosphere. Analysis of meteorites known to have originated on Mars and the moon by scientists at the Johnson Space Center shows that the outer edges of such rocks melt in the fiery heat of entry. But the interiors stay frigid, chilled to the icy temperatures of interplanetary space, according to planetary geologist Carlton Allen at the JSC. A study published in the journal *Nature* in 2000 showed that the controversial potato-size Martian meteorite Allen Hills never reached a temperature greater than 100° F during its fiery descent through Earth's atmosphere. Any microbes inside the rock would have been safe in their miniature caves. Several tons of unsterilized Mars fall on Earth each year. Although the reverse trip is less likely, computer models suggest that at least a few pounds of Earth makes it to Mars in any given year. In the words of Chris McKay, an astrobiologist at NASA's Ames Research Center who has collected extremophiles from Carlsbad Caverns National Park, "The

planets are always swapping spit."

Hoover, considered a leading expert on the tiny, plantlike creatures called diatoms, has become an outspoken panspermia advocate. He has claimed to spot evidence of space-faring microbes in carbonaceous chondrites, meteorites believed to have originated in the asteroid belt. While those claims remain hotly contested, Hoover has found greater support from the world scientific community for the extremophiles he gathers from remote locations. In 1999, he successfully grew a 40,000-year-old moss recovered from an Alaskan glacier. In January 2000, Hoover joined Apollo 13 mission commander Jim Lovell and seven others in a hunt for exotic microbes in the ice and meteorites of Antarctica.

"The extremophiles in the Antarctic glaciers and permafrost represent analogs for microbes that may someday be found in the permafrost or ice caps of Mars or on other icy bodies of the solar system," says Hoover. "Ancient microbes can remain viable while frozen for thousands or millions of years, resuming metabolic activity after freezing. These bugs are out there. All we have to do is find them."

investigation that her family still jokes about, seeking to answer the question of how food that went into her came out so utterly changed. Excited to encourage an interest in science, her grandfather sat Hazel on his knee and attempted to explain. In the process, he began what she later termed "my lifelong fascination for how 'in' things become 'out' things."

Her first exposure to science was an elementary school biology class. A teacher named Martin Upson introduced the enthusiastic but easily distracted girl to the concept of an environment: a place where creatures live and rely on each other to maintain a balanced ecosystem. Hazel's project was to go to a nearby pond, fish out a pot of goo, bring it back to the classroom, and look at it under the microscope. She then wrote a long report about the animals that lived in the pond, how they interacted with each other and with the ecosystem. She included drawings of the creatures she found, their scientific names, and a description of their life-styles. In the evenings she diligently colored in the drawings from class.

"I was hooked," she later recalled. "At 11 years old I was a fledgling scientist, a passion that has remained the driving force of my life ever since."

But when she was 16, another enthusiastic teacher, Jim Moon, pointed Hazel's natural curiosity in a new direction. She had a choice of physical education courses that year: play hockey, play tennis, or go caving. Moon was leading a caving trip to the Mendip hills. She was the only girl in a group of teenage boys, all of whom laughed when she slid 20 feet down a mud slope the rest had climbed with no trouble. When one bet that she'd never venture into a cave again, she vowed not only to return but to become a better caver than any of them. She had found what would become her life's second great passion.

By the age of 18, she had joined the Wessex Cave Club, learning how to negotiate pits on increasingly demanding trips. While studying biology at the University of the West of England, Hazel supported herself by working as a barmaid in a popular Bristol disco. After Saturdays spent underground, she would spend Saturday nights drawing pints, her light brown hair teased so tall "it added close to a foot to my height." She would sometimes grab spoons and drum behind the bar to songs she liked, prompting the deejay to encourage the crowd to "give a big hand for H." Her friends began to call her "H," a nickname that stuck.

After earning a bachelor of science in biology, with honors, she received a fellowship to a doctoral program in microbiology at the University of Colorado, and took off for a country where she knew no one. Luckily, she knew how to find cavers. Within days of arriving, she had arranged to go on her first trip with members of the Colorado Grotto (as most U.S. local caving clubs are called) of the National Speleological Society. H soon became a frequent explorer and surveyor in Wind Cave National Park, traveling sometimes for 20 hours underground within the complex, multilevel cave system. Even as she was learning delicate microbiological techniques, she developed her skill in cave cartography, drafting dozens of award-winning cave maps. After earning her Ph.D. in 1997 with a dissertation on *Pseudomonas aeroginosa*, the bacterium that kills patients suffering from cystic fibrosis, Hazel became an instructor in the surgical department of the University of Colorado Health Sciences Center and chair of both the Colorado Grotto and the larger Rocky Mountain region cave survey.

And she still fumes when she recalls the boys who laughed at her on that first trip—although she readily admits that all are now good caving friends.

Nancy Holler was caving before she was born.

Her mother went caving when seven months pregnant. After Nancy was born, her parents and two brothers founded the

Flittermouse Grotto, a local chapter of the NSS, which requires at least five members to start a new club. Nancy can recall, vaguely, being carried through caves in a backpack on her father's back. At an early age, she learned to appreciate caves not only for their beauty but also for their scientific resources. While growing up, she helped her family conduct cave biology inventories, bat counts, water studies using dye traces, archaeological inventories, surveys, and clean-up trips.

As the smallest member of the family, Nancy's parents would often have her squeeze into unexplored cave passages where no one else could fit. When she was five years old, hiking back to the family car from a trip with her father and brother Chris to Warrior Mountain Cave, the three discovered a virgin cave. A tight opening between boulders beckoned. Ever eager, Nancy volunteered to crawl in first. A mass of crickets, common to limestone caves the world over, clung to the ceiling in an undulating brown mat. Her helmet touched them as she crawled—and hundreds rained down on her, sliding into her clothing.

"It totally wigged me out," she says now. "The little guys always jump on me, never on anyone else. It's like they know my past."

As a teenager, Nancy began writing and presenting scientific papers at the annual NSS conventions. She was twice awarded for the best paper presented by a member under the age of 25. In her late teens, she became more independent and began venturing regularly to TAG, where she quickly made new caving friends.

In the 1970s, improvements in gear and increasing public interest in ecology had combined to swell the ranks of cave clubs and annual caver gatherings. By the 1980s, cavers increasingly traveled to distant parts of the globe to explore previously unchecked limestone regions. Having seen miles of actual frontier recede before their headlamps, a few of the cavers with whom Nancy socialized had adopted the swagger of the conquistador, others the honest and uncouth humor of the mountain trapper. They partied hearty. Like other pioneers, they also tended to be casually magnanimous: without thinking much about it, a dozen would help a friend build a garage or move across the state. It was considered perfectly normal for five or ten strangers to drop in unannounced to spend the night on the floor—or at worst in the yard—of a caver who happened live en route to an expedition or gathering.

In the fall of 1992, Nancy, 21, arrived late to the annual TAG Fall Cave-In, a huge gathering of cavers. The only spot left to pitch her tent was in a mud wallow beside a row of portable toilets. A strong young caver named Brent Aulenbach was stuck there too. As often happens in the insular community to which both belonged, they soon learned that they knew many of the same explorers. Brent and Nancy caved together several times that weekend. Six months later, amid a March blizzard in North Carolina, she got a letter from Brent asking whether she planned on attending SERA—the spring caver gathering of the Southeast Regional Association of the NSS. Once again, the two went on several caving trips together, this time to the deep pits of TAG. Nancy invited Brent to ride along with her that summer to the NSS national convention in Oregon. On July 30, 1993, at the Million Dollar Cowboy Bar in Jackson Hole, Wyoming, she realized that she was falling in love.

In 1994, she accepted a one-year internship at Jewel Cave National Monument in South Dakota. There she helped explore and survey Wind and Jewel Caves, conduct and analyze water samples, and perform bat counts. Once her internship was complete, Nancy moved to TAG to be closer to Brent, her friends, and her favorite caves. She landed a job as a schoolteacher in suburban Atlanta, and in January 1996, while climbing up separate ropes out of Neversink, a classic TAG pit 165 feet deep, Brent asked Nancy to marry him. Shortly after the wedding, Nancy was elected to the NSS Board of Governors.

She and Brent continued to cave every possible weekend.

In 1998 John Scheltens, a well-known explorer of Wind Cave and past president of the National Speleological Society, became an advisor to MacGillivray Freeman Films. He suggested two exemplary cavers as possible focus characters for a planned IMAX theater film on caving: Nancy Holler Aulenbach and Hazel Barton. Joined by a largely French-speaking team Nancy and Hazel bonded quickly, in part because both understood very little of what was said in French. In addition to the Lambertons—Janot, Mael, and Janine, Janot's wife—the team included Luc Moreau, a glaciologist; Karime Dhamane, a French chef hired to cook for the expedition; and a local Greenland caver named Kim Peterson. Among the American filmmakers already with the French when Nancy and Hazel had arrived were two riggers, four cameramen and assistants, a production coordinator, and the director and head of the expedition, filmmaker Steve Judson.

Hazel and Janot worked their way along the far side of the river canyon while Nancy and Mael set up a rope on a nearby ice wall. They both resolved that no harm could come if they would at least experiment with each other's gear. International tensions eased a great deal when the cavers discovered things they liked about one another's climbing apparatus. After seeing how well the Petzl clung to ice-coated ropes, she agreed to use one during the expedition. At the same time, Mael recognized Nancy's proficiency on rope and no longer criticized her gear. Later the entire expedition sat clustered in the dining tent, toasting one another and telling caving stories in three languages. Eventually Nancy pulled out Pass the Pigs, a sort of dice game that required throwing plastic porkers. They played long into the arctic night. Outside the northern lights flashed, shooting reds, greens, and yellows overhead like a New York disco. The temperature slipped at last down below zero, and kept falling.

During the expedition's unusually warm fall weather, afternoon heat would melt small pools in the ice near camp that would freeze over again each evening. Here, team members improvise an ice rink—sans skates—at the end of a long day spent collecting microbes underground.

43

CHAPTER 3

IN SEARCH OF NEW LIFE

The cold arrived via an 80-knot wind straight from the pole. Tents wobbled and rattled all night; Nancy and Hazel didn't think theirs would make it through intact. They shivered through several layers of clothing, boots, hats, gloves, and doubled sleeping bags, as if they slept on bare ice. Condensation from their breath froze on the bags, coat-

ing them in a thin white layer. At one point Nancy looked at her watch and discovered that the cold had affected its liquid-crystal display: the seconds crept along at about a tenth of normal speed. With the tent shuddering like a trapped beast, she watched the frozen seconds—38...39...40—tick away in slow motion. ■ When morning came they were amazed to find that only one tent had come down, its hapless resident moving to the dining tent. As the sun crept in its low arc through the wind-cleared sky, the temperature once more began to rise. After breakfast, Janot, Mael, and a few members of the team hiked to the nearby mouth of the moulin to gauge the possibility of gaining entrance. ■ "Not today, but maybe tomorrow," Mael pronounced at the edge of the abyss. ■ The day before, meltwater had backed up nearly to the rim of what just a week earlier had been a pit perhaps 600 feet deep. As the temperature dropped overnight, the surface of the plugged drain had frozen over. The force of the stream had slowed, water gradually reced-ing below. Now a few inches of transparent ice covered deep blackness.

Above: Icicles in the moulins could form deadly projectiles. Right: Occasionally safer, naturally smooth shafts present them-selves, as caver Diana Gietl discovered on Janot's 1994 expedition.

With thousands of pounds of film equipment to move across frozen rivers and crevasses, the filmmakers continually had to rely on their ingenuity. Here, crew members Phillipe Dufieux,(left), Karim Dahmane, Brad Ohlund, and Kim Peterson use a makeshift sled to haul gear.

"I don't think you want to go skating there," Nancy said.

Hazel nodded. "Well, then, I guess it's back to the traverse." The group agreed that science and exploration would have to wait for another day or two of cold weather. Filmmaking became the order of the day.

The previous afternoon, Hazel had rappelled in the midst of her traverse to examine the river close-up, dropping to within a few feet of it. Dave Shultz, the team's chief rigger, had come up with a way to re-create the rappel with the heavy IMAX Mark II camera and film, giving audiences the same gut-wrenching view H enjoyed. It was a dangerous and complicated shot, requiring a motor-driven winch and several people coordinating ropes and safety brakes.

Gordon and his camera were harnessed together like a rescuer and accident victim. Looking through the viewfinder proved difficult with his climbing helmet in place, so he opted to leave it on the ice. He would be hanging over an open space, where, he reasoned, falling ice was unlikely to be a concern.

While Shultz ran the winch, French caver Jacques guided the rope holding Gordon and his gear onto a large capstan. The friction provided by wrapping rope around the metal post would help control the rate of descent and reduce strain on the winch during ascent.

Hazel volunteered to man the emergency brake, which was an anchored Gibbs ascender. Little more than a metal cam inside an aluminum shell, the Gibbs allowed rope to slide freely through unless "loaded" by weight placed on the cam. The device normally attached a climber's foot or chest to a fixed, free-hanging rope, and was alternately loaded by the climber's weight and released whenever the climber took a step upward. Cavers call it rope-walking; with two or three Gibbs in combination, they can "walk" up deep pits as easily as climbing stairs. But to use the Gibbs as a safety brake for a load on a moving rope—in this case, Gordon and a specially designed IMAX camera—required the person manning the brake to hold the cam open. In an emergency, Hazel would have to let go of the Gibbs so that it would lock onto the moving line and keep Gordon from plunging into the icy stream.

The crew working the system remained a safe distance from the edge, each of them securely anchored to the ice. Shortly after Gordon gave the signal to be lowered, he was out of sight from those who controlled his rope. Other members of the team stood closer to the lip, relaying his commands.

Despite the complicated setup, the descent went exactly as planned. Gordon got his shot, to the delight of director Steve Judson and others anchored to the cliff. But as Dave winched Gordon back up the rope, Jacques began having trouble keeping up at the capstan. A mound of slack accumulated. Abruptly the rope—and Gordon—began sliding back toward the abyss.

Steve and his crew members at the lip were horrified to see Gordon vanish from sight, plunging toward the river at increasing speed.

Hazel, trying to decipher Jacque's French, took a second or two to realize what was happening. As rope sang through the Gibbs she knew she had let go: she jerked her hand into the air—the Gibbs caught; the line stretched instantly taut. A loud crash arose from below, followed by muffled cursing.

Hazel had stopped the uncontrolled plunge before Gordon hit bottom in a potentially fatal fall, but the sudden tension on the line slammed him into one of the canyon walls. His head and shoulder absorbed the blow as he tried to protect the camera. The front element of the Mark II's 30-mm lens was knocked off and plunged into the river below, but he saw that the camera and its film were intact. When he put his hand to his head, he saw blood.

When the crew hauled Gordon at last to the surface, he was bleeding from a gash in his forehead but was able to walk to camp unaided. The expedition's doctor diagnosed a mild concussion. It took five stitches to sew up the wound.

Everyone was shaken by the incident. All knew that the most dangerous work was yet to come, in the caves below.

Gordon was sore and stiff but ready to go back to work after a day of rest; Hazel, who blamed herself for the injury, was at last cheered up after a comical ice-skating session in camp. And Nancy reminded H that she had almost certainly saved Gordon's life, that the accident—as is so often the case with any highly technical and dangerous effort—resulted not from any one error, but from the snowball effect of a series of otherwise minor mistakes combined with miscommunication. In the end, the safety backup had worked exactly as it was supposed to: tragedy had been averted.

Hazel had another reason to be slightly upbeat the next day: the cold had maintained, and Janot had declared that they could safely enter their first cave, Minnik II, a horizontal tunnel with a second vertical "skylight" entrance shortly inside. Hazel would be able to collect her first cave specimens. She had already sampled some tardigrades from algal mats of one of the small ponds near camp. The important question Hazel hoped to answer was whether these tiny single-celled animals could survive thousands of years of cold storage in the deep ice.

Continued on page 53

Cameraman Gordon Brown is strapped into a custom camera harness that will support him as he hangs out over a deep crevasse to film a dramatic shot. The filmmakers agreed that the rigging required for shooting in ice caves was among the most complex they had ever experienced.

Luc Moreau leads Nancy and Hazel over the jagged fallen roof of a horizontal ice cave as they work their way toward the entrance of Minnik II. Two days before, the team had heard the rumbling collapse from camp, nearly a mile away.

OF TEDDY BEARS AND TARDIGRADES

BY HAZEL BARTON, PH.D.

As I peered into the microscope, Philippe Bourseiller breathed broken English into my ear. "When you see one, you will know. You will fall in love, 'e is so beautiful."

I continued to scan the slide in search of my prey, the tardigrade. Janot Lamberton had spoken much of them since we'd joined his expedition in Greenland. My interest was the unique bacteria living here, but Janot insisted on taking me to a patch of ice goo; there we would find some tardigrades—animals, not bacteria, yet likewise visible only through a microscope. As a biologist, I am intrigued by anything different, and was more than happy to hunt for these tiny organisms that eke out an existence on the frozen tundra. After two days of looking through the sample for the elusive beasts, I was beginning to think they were a figment of someone's imagination.

Janot told me to look for "teddy bears." I had no clue what he meant. As Nancy and I worked in the science tent, Phillippe had joined us to prepare camera gear. Also fond of the tardigrade, he offered to join the search. To help, he drew a rough sketch of a blobby looking creature, with what resembled a bear's head at one end. With the rudimentary figure in hand, I continued to scan the one under the microscope.

"Oooh! Is this one?"

I leaned back and allowed Philippe to peer into the microscope. "Yes. It is one. See...he waves at you."

One of the appendages of this tardigrade was bent backward in a mock wave. I finally began to understand what Janot had been saying about "teddy bears." The little legs of the wee beasty were bowed outward, and with tiny eyes and structures

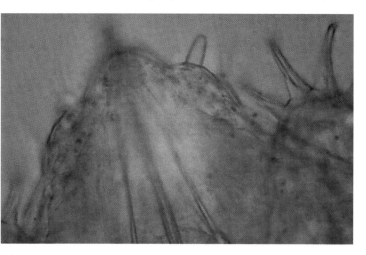

The red blobs, "eye spots," in this 300X magnification are light sensitive.

around its head, the tardigrade did indeed look like a teddy bear walking on all fours, complete with a big smiley face.

As small as a tenth of a millimeter in length, the tardigrade was one of the first creatures to be seen by Leeuwenhoek when he invented the microscope over 300 years ago. Upon discovery, they were given names such as "water bear" and "moss piglet," to describe their animal-like appearance. However, it wasn't until the 18th century that the organisms were named tardigrades, from the Latin for "slow"—tardi—and "walker"—grado—to describe how they lumbered around.

The evolutionary history of the tardigrade remains unclear, until molecular techniques can resolve its past. Presently, they are thought be part of the ancestral lineage of present-day insects, a family of creatures that found their niche early in the biology of our planet and have lived there relatively unchanged for millions of years. Indeed, tardigrades have been found in ancient amber, showing very little change in their body architecture for the last 65 million years.

What makes these creatures so interesting is their ability to survive some of the harshest environments on earth. Unlike bacteria, which are single-celled creatures (essentially a bag of enzymes), the tardigrades are multicelled beings—with rudimentary blood, immune systems, and even brains. As complex organisms, they have adapted to live in deserts, hot springs, and even the frozen tundra of the Arctic. Here, with our expensive and advanced survival gear, humans may only visit briefly, while the tardigrades can live for 100 years in this frozen waste. The tardigrades were the only other animals with which we shared the ice. Even the hardy polar

The tardigrade's mouth is a thin tube designed to suck algae. Unique in nature, the tiny animal can survive long-term freezing.

bear is confined to the very edges of Greenland, where water and food lie in relative abundance.

So how do they do it?—How can the tardigrades survive here? They can't avoid freezing. Instead, tardigrades have evolved to tolerate repeated freezing.

When our tissues freeze, ice crystals form from the water in, and surrounding, our cells. Inevitably all the water is incorporated into one or two large ice crystals, which push through and rupture the membranes of our cells. In addition, as the water freezes, it is sequestered from the liquid center of our cells. As the water is lost, our proteins and cell membranes are damaged; they need to interact with water molecules to remain in the correct shape. The result is severe damage and cell death. If enough damage occurs, areas of frostbite develop and the tissue dies.

The tardigrade avoids this kind of cell damage. In its "blood" are special molecules, called ice-nucleating agents, which actually encourage lots of ice-crystal growth. As a result, there isn't enough water available for any one crystal to grow particularly large and damage the cells. The tardigrade also produces its own antifreeze, which slows the freezing process, and makes a protectant to stabilize the proteins and membranes of its cells as they are dehydrated. With these mechanisms, a tardigrade can be frozen solid, with no measurable metabolic activity—a state termed cryptobiosis—and brought back to life upon thawing.

Tardigrades offer a potentially huge scientific benefit to humanity—if we can harness the mechanisms of cryptobiosis. One of the biggest problems for medical research is the ability to study cellular activities in culture. A culture is a homogenous population that allows us to determine how cells grow and divide or the effectiveness of certain drugs on those cells. However, some cells do not grow well in culture and must be freshly isolated—a very laborious process dependent on someone's generosity to donate a healthy tissue sample. If we could somehow maintain large volumes of these isolated cells by placing them in the same cryptobiosis by which the tardigrade survives the Arctic winter, then it would be possible to maintain a source of cells for years.

Very soon news of our find had made it around camp. Everyone, including the IMAX-camera crew, stopped by the science tent to peer down the microscope and look at our waving friend. All the time Janot stood by, like a proud father, pleased to introduce our only neighbor in this frozen landscape.

Modern microbiologists have developed two methods for extracting new life from extreme environments. The traditional, if more difficult, choice is to "bring 'em back alive." The first two centuries of microbial knowledge came from live cultures grown in the laboratory. Thus the microbes that lived at room temperature, liked oxygen, and could eat more or less the same sort of food as humans were the first to be studied comprehensively. Only gradually did microbiologists begin to realize that there were many other species of microbes-perhaps the majority of species-that would die at what we consider "normal" temperatures, or in the presence of oxygen, or when fed anything more rich than a chip of rock. These extremophiles were extraordinarily adapted to environments very unlike the average science lab, and thus new collection methods, culture media, and isolation chambers had to be invented in order to study them.

Beginning in the late 1970s, microbiologists began developing another approach to understanding microbes that died when examined by traditional means: the new science of genetics. Using recombinant DNA and RNA techniques, Carl Woese of the University of Illinois began extracting and amplifying genetic material taken from microbes in a variety of environments, from black smokers—volcanic fountains of hot water, minerals, and microbes on the ocean floor—to pipes that carried radioactive water through nuclear plants. By looking at a particular gene that mutated at a known rate and was shared by all living things, he could learn a great deal from a single cell—even if it was dead, even if it had never been cultured in any laboratory. By comparing the family relationships, or phylogeny, of many such organisms, Woese was able to construct a phylogenetic Tree of Life. Woese's work was further advanced by caver and microbiologist Norm Pace, who created faster, more accurate methods of analyzing the genetic relationships of novel organisms. In the 1990s Pace's lab had shown conclusively that the vast majority of life and genetic diversity on Earth was microbial and that fewer than two percent of all microbial species had ever been cultured or studied.

Janot (left) leads Hazel (right) gingerly over an ice ledge deep within Minnik II, while behind them Nancy (top) waits on another ledge as Mael checks out a descent route below her. Although such ledges appeared sound, they would sometimes shift with a loud crack, threatening to send team members and IMAX camera gear tumbling into the depths.

Above: Hazel chips through fresh ice to get at the preserved algae pools in a dark band deposited in an Arctic summer long past. Below: In the science tent, Mael helps Hazel spot tardigrades swimming in a sample.

After examining the entrance area, she and Hazel hiked with Mael and Janot to the skylight entrance to rappel into a deeper section of the cave. They discovered that the ropes they had rigged only an hour before were encased in several inches of ice formed by spray from the small stream flowing into the skylight. After re-rigging the drop, the cavers descended, using their crampons to walk down a smooth face of flowing ice. Nancy couldn't get over how much the ice resembled the cave mineral called flowstone; if this were a limestone cave back home, she would have taken great pains to avoid touching such a delicate calcite formation, let alone bouncing down and marring it with crampons. A single caver's passing could scar the formation in a way that would remain visible for thousands of years. Yet in the transitory realm of the ice cave, Nancy knew that all signs of her passage would be erased long before she returned home to Georgia.

In the deepest reaches of Minnik II, Hazel was able to pull out the ice ax and chop away at a portion of ice wall that contained the same dark particles in which she had found tardigrades on the surface. This horizontal cave was not as deep—and thus not as old—as the group hoped to get in Malik, but it was a start. After emerging from the cave, Hazel placed her samples in a large metal canister containing liquid nitrogen. While there was no need to keep microbes alive for successful genetic analysis, it was important to preserve their DNA, which would begin to degrade within hours if kept at any temperature above freezing. Her samples had to remain well below freezing if any useful science was to come from them.

Later that day, Janot and Mael were able for the first time in weeks to enter Malik and examine the results of the flood. Loose ice clung to the walls everywhere: the two worked down, using their hammers to dislodge massive chunks. Although the waterfall in the pit was only a tiny fraction of its size during the warm days that had passed, it was creating enough spray to glaze the ropes in a thick coating of ice, making it hard for ascenders to grip.

The two exited the pit, declaring that it was still unsafe for the rest of the group to enter. Science and a new depth record would have to wait.

It was to the Pace Lab that Hazel hoped to bring tardigrades and other microbes from below the ice cap, in order to use chemical and genetic methods to deduce their recipe for suspended animation.

The entrance to Minnik II was a snow bridge, spanning the in-rushing stream. Two days before Nancy and Hazel had arrived, a huge part of the bridge had collapsed, destroying a long section of cave passage in an instant. The French had heard the roar of the collapse in camp, a quarter mile away. Fortunately, Janot and Mael knew what signs to look for and declared that the surviving entrance and passage was now frozen solid, and safe. Nancy worked her way to the opening, 60 feet tall by 150 feet wide, marveling at the incredible blue aura which soon engulfed her.

LIFE UNDERGROUND: A NEW VIEW

BY NORMAN R. PACE, PH.D.

Microorganisms receive little attention in our general texts of biology, are largely ignored by most professional biologists, and are virtually unknown to the public except in the contexts of disease and rot. Yet the workings of the biosphere depend absolutely on the activities of the microbial world. In just a handful of soil from Earth's surface are billions of microorganisms, of so many different types that accurate numbers still remain unknown—and it appears that the greatest portion of this microbial universe lives beneath Earth's crust, within the subterranean conduits and habitats of the planet. Microbial biology seems to loom large in the sustenance of life on, and within, our planet and has much to tell us about how life on Earth evolved.

Microorganisms are tiny, individually invisible to the eye. The existence of microbial life was recognized only relatively recently in history, with Leeuwenhoek's invention of the microscope. Even under powerful microscopes, the simple structures of most microorganisms prevents their classification by morphology, through which large organisms have always been classified. However, with the recent development of methods for classifying organisms by their DNA sequences, it has become possible to determine the evolutionary relationships that connect all of life. These results have allowed us to diagram a universal evolutionary tree—a graphical representation and methodology

that has profoundly changed our view of the history of life on Earth.

Before the development of the molecular methods that allow us to examine DNA, evolution of life was interpreted in the context of five kingdoms: animals, plants, fungi, protists (protozoa), and monera (bacteria). The relationships between these kingdoms were largely speculation, however, and had never been tested—until Carl Woese of the University of Illinois formulated a more objective view of evolution in 1977 by comparing DNA from the ribosomal RNA (rRNA) gene of many organisms. From these comparisons, he was able to reconstruct a history of life by describing connections between all organisms, including microbes. His discovery proved the five-kingdom concept of classification to be fundamentally incorrect.

The rRNA gene is a biological structure that changes very slowly, making it possible to map the rRNA gene's evolutionary changes. Woese established the now recognized three-domains framework of evolutionary descent: the Eucarya (eukaryotes, containing a nuclear membrane), Bacteria (initially called eubacteria), and Archaea (initially called archaebacteria). This current evolutionary tree, based on rRNA gene sequences, calculates differences in the rRNA gene between each organism and plots a map of "evolutionary distance," which considers only changes in gene sequences.

Until this point, the evolutionary diversity of life on Earth was thought to be composed of multicellular forms,

animals, plants, and fungi. Biodiversity was articulated in terms of large organisms; using rRNA gene sequencing, it is now possible to describe our planet's microbial diversity—but perhaps its greatest application is identifying evolutionarily primitive organisms, which clearly cannot be cultivated today. So far, our understanding establishes that the most ancient organisms on Earth apparently were those that used hydrogen as an energy source and were thermophilic, a finding consistent with current theories: Earth's first life-forms lived, fed, and reproduced in a geothermal setting at a high temperature.

Our understanding of life's diversity is thus just beginning. Most of our recent knowledge has portrayed only a part of the global distribution of life— the part that has been immediately dependent on either the harvesting of sunlight or the metabolism of the products of photosynthesis. But our current Tree of Life indicates that photosynthetic metabolism evolved long after inorganic energy metabolism, which existed during a very early stage in Earth's history when the biosphere was based on hydrogen and other inorganic compounds. Today a similar biosphere extends miles into the Earth's crust, an essentially unknown realm—indicating perhaps that much of the living matter on Earth is isolated from light, thriving within the planet's cavernous terrain We have begun to understand that the biological diversity of life-forms is likely to be as pervasive in occurrence underground as it appears to be on Earth's surface.

CHAPTER 4

OUT OF THE FALL ZONE

The vast Malik Moulin remained unstable, showering fountains of ice upon Janot each time he and Mael attempted to examine it. Frustrated, the team began to turn its attention elsewhere. ■ After the near disaster on the traverse, no one was taking even slight chances with safety. Kim, the team's Greenlander fireman, had been appointed safety officer, and he personally checked each piece of gear before anyone descended into the ice. Janot was the first person into each cave, carefully studying the walls and roof for signs of potential collapse. All wore helmets when on rope. There had been no more climbing mishaps, but while testing a small gasoline-powered winch, Mael had caught his thumb in the gears, ripping off the nail and part of the flesh.

Above: Moist air moving through the caves causes ice crystals to form near entrances, at times in the space of hours encasing any gear cavers have left near the surface. Right: Janot Lamberton hammers at a loose shield of ice before descending beneath it.

■ "If your skin weren't so thick, you'd have lost the thumb," the team doctor said as he bandaged the wound. ■ Nancy and Hazel were becoming more adept at maneuvering on the ice. In camp, they had learned to use their axes to nightly rough up the ice outside their tent, so that they would have sure footing when they stepped out in the morning. Underground, the differences in rope work required for ice—as opposed to terrestrial caves—were becoming second nature to both. ■ When rigging pits in TAG, Nancy was accustomed to long, "free" rappels, where the rope seldom touched a

wall before hitting bottom. In Greenland, the danger of falling ice required rigging as close to the sides as possible, with frequent anchors called rebelays fixed to overhangs in order to move the rope away from the fall zone. Nancy was glad she had practiced crossing rebelays while rappelling in Georgia. She was able to impress at least one of the French cavers with her speed and grace on rope as she crossed a rebelay while dropping into Minnik I. Yet the French still argued with her over the Texas-style rig she used on ascent, a system they had never seen.

Malik remained out of reach, so a few days after injuring his thumb, Mael joined his father in exploring a second vertical moulin, Minnik II, part of the same stream system flowing through Minnik I. The depth potential of Minnik II was not as great as Malik's; but at well over 350 feet deep, the pit would allow Hazel and Luc scientific access

to older ice than the expedition had yet encountered. What's more, about halfway down, Janot spotted a recessed area atop a huge ice block where the crew could rig a filming platform protected from the constant spray of shards from above.

Hazel and Nancy had noticed another marked difference between caving in rock and in ice: the team's approach to mapping. Even for the most experienced explorers, caves can be confusingly complex. Passages branch continually in three dimensions. So-called straight tunnels can undulate like serpents. Nowhere in a cave are there any of the familiar reference points— sun, moon, trees, fields—by which humans normally find their way on the surface. Thus virtually all cavers who explore virgin passages become accomplished surveyors, carefully recording distance, elevation, passage shape and size as they move through a new find for the first time.

A few cavers, and Hazel Barton is one, have also become talented mapmakers, drafting the survey data into large maps that detail a cave's plan, profile, geology, and major levels or sections. Often cavers will not recognize that a main passage follows a particular fault or rock layer until they study the finished map. The map becomes the caver's most useful tool for finding new passages within a known cave.

The problem with mapping one's way through a cave is that it can be very time-consuming. Taking initial survey data can slow a team down, frustrating anyone staring at a yawning blackness ahead. But by far the most time-intensive part of mapping comes on the surface. With a fairly small, straight stream cave, plotting the data and drawing a finished map can take up to a few weeks. With very large cave systems, such as those in Mammoth Cave, Carlsbad Caverns, and Wind Cave National Parks, survey may require many separate maps for different sections, with each map representing decades of continual revision by several hundred volunteers.

Ice caves forbid such mapping. Of course, sooner or later all caves are transient; give it 20 or 30 million years and the limestone now enclosing Mammoth Cave will wash to the sea. But caves of ice are transient on a human scale: no single cavern survives the length of time it would take to properly survey it. Anyone who saw the finished film of the expedition would be viewing places that no longer existed.

After years of studying glacial caves Luc Moreau had come up with a streamlined survey method, which he taught to Nancy, focusing on key changes in the living glacier that held the cave. Except for using a tape to measure passage depth, he ignored the types of measurements most commonly taken in cave surveys: bearing, inclination, and distance. Instead, Luc concentrated on getting very precise measurements of passage width, in order to gauge the weight and force of the ice surrounding the cave.

He would place wooden pegs into opposite sides of a shaft and affix to their tips electronic devices that could record the exact distance between them. Each of the deep moulins was wider just below the top than at bottom. The meltwater shooting from the *bedier*s, or surface canyons, would at first carve a bell-shaped opening at the back of the pit. But the great weight and pressure of the ice gradually squeezed the pit's pliable sides together at depth, until the water vanished into tight conduits impassable by human explorers. In the limestone pits of TAG, the exact opposite often occurs, with water eroding an enormous cavern at the base of a shaft whose top is barely wide enough for cavers to squeeze hips and shoulders through.

By using precise electronics and leaving the pegs that held them in place during an expedition, Luc could observe even a few centimeters of change from one day to the next. After plotting the changes in distance over depth and time and recording the data

Luc Moreau demonstrates his method for mapping the daily changes in ice caves to Nancy. Pegs hammered into opposing walls contain electronic measuring devices that record the distance between them and measure daily changes as the walls slowly squeeze together under the massive weight of the ice above.

in a computer, he could begin to form a clear picture of the behavior and composition of the surrounding glacier.

Hazel, who had chaired annual cave-mapping contests hosted by the National Speleological Society, wished that she could

Nancy uses a device called a bobbin to rappel into an ice cave. Unlike other descent tools, such as figure eights or the rappel racks favored by limestone cavers, bobbins resist slipping on ice-coated ropes.

representing past summer pools. In the deep moulins, the constant spray from inrushing water would coat the ancient walls in several inches of transparent new ice. For a microbiologist, this was a good thing: by sampling from several inches beyond the border of new ice and old, Hazel could be reasonably certain that her samples had not been contaminated by contemporary organisms. The trick was to chop through the ice, then use sterile tools to quickly remove and isolate the sample. This would have been difficult enough in and of itself, but the twin needs to dodge falling ice bombs and record the work for a camera crew made the collections downright sporting.

Much of the first day at Minnik II was consumed with the delicate task of mounting a platform for the camera on the loose wall of the cave. In order for the film crew, the support team at the surface, and the various cavers and scientists to communicate during the complicated setup, everyone carried portable radios. Hazel, Nancy, and others who would appear on film had been fitted with passive microphones wired into the radio system. During the long, cold wait while Gordon Brown and others worked on the platform, Hazel and Nancy hiked back to camp to warm up, their radios on so they could be summoned back to the pit when needed.

draft a precise map of just one of the ice caves explored by the expedition, then continue to resurvey it as it changed over time. Ultimately, such data could be used to create a three-dimensional animation of the complete life cycle of an ice cave. But such a project would take far more manpower and time than this expedition could spare. So while Nancy assisted Luc with his width measurements, Hazel focused on microbial collection.

Minnik II seemed a likely shaft for providing ancient tardigrades and perhaps new species as well. Like other caves the team had examined, it sides were marked by the dark lines

Back in Minnik II, small and large chunks of ice popped out without warning, exploding with noises that reverberated through the confined pit. Each time an anchor was pounded into the ice, it seemed as if the sound itself could send the walls tumbling like a house of cards. Gordon's brother, Mike, an experienced mountain climber, was assisting him on the small platform, and kept muttering over the radio that he'd never seen such a dicey-looking place. The first shots taken from the platform show blue blurs streaking past the camera, meteors of ice flying just outside the camera's protective alcove.

ON ROPE

BY NANCY HOLLER AULENBACH

Humans have always emulated animals—flying through the sky like a bird, speeding over the land like a cheetah, swimming through the seas like a fish. Vertical cavers like me emulate the spider, its grace and ease on silken threads.

My first brush with vertical caving began seven months in utero as my mom belayed my dad while he did a traverse and descent into a Tennessee pit. Later my baby pack became my first seat harness. Mom and Dad inventively secured it on rope and hoisted me, along with my two older brothers, up and down small drops. Rigging a rope in the old oak tree was the Holler family's idea of a backyard swing. I recall hanging from my harness, being pulled gently to and fro. Graduating when I was older to rigging ropes hung off of our second-story deck, I practiced changeovers from rappelling to ascending, all under the watchful eyes of my parents. Now, living in TAG, a region full of vertical caves where Tennessee, Alabama, and Georgia meet, I can be found on rope almost every weekend. One day, dangling 150 feet from the floor of a cave on parallel ropes, my caving husband proposed to me and swung over to place the ring on my finger!

Needless to say, I feel very comfortable on rope.

Vertical caving has certainly changed a lot over the years. Early people used vines and tree limbs to reach the depths. When caving became a scientific endeavor in Europe, cavers would allow themselves to be lowered and hoisted by the aid of their fellow explorers. Cumbersome and heavy rope ladders and more recently cable ladders were the norm for long descents, but limited how deep explorers could go.

It has often been said that necessity is the mother of invention. In 1952, Bill Cuddington, an American caver, was unable to find others to hoist or belay him in a Virginia pit. He had body-rappelled many times before, so he could easily get down the rope. Getting up, however, was the trick. He had read about Prusik knots in an old mountaineering book; by tying three short lengths of cord to a larger rope—one for each foot and one for the climber's chest—a person could "slide" up a rope without slipping, scooting the knots upward incrementally. After practicing tying the knots onto a stout rope, he was ready to give the method a field test; that year Bill became the first to successfully rappel and climb a pit using the single-rope technique that is still used today.

Metal ascenders have now taken the place of Prusik cord; body rappels have given way to mechanical descenders such as bobbins and racks. However, it is important to learn the basics and practice regularly so you can get yourself out of almost any jam. It has become far too common for cavers with little training to get stuck on rope and not know how to perform a self-rescue through improvisation. Learning how to cut off the tail end of your rope, for example, to fashion Prusik slings is invaluable knowledge. Body rappels, though quite uncomfortable, can come in handy too. For years now, I have enjoyed being a part of the National Cave Rescue Commission. Although today's sophisticated rescue equipment and techniques are available, often the best solution to a problem underground is for the those in distress to try and handle matters themselves, before a full-blown rescue is called. For this reason, a big part of the vertical training done by American caving clubs, or grottos, is to have the trainee simulate emergency situations in a benign environment.

Cavers have always been a proud and inventive group of explorers. Climbing techniques and equipment are as variable as cavers themselves; and we all make proud attempts to convince others that our system is superior. This friendly debate is nothing new. In 1946, Robert deJoly, the president of the French Speleological Society wrote a spirited letter to the National Speleological Society founder, Bill Stephenson, to persuade Bill that deJoly's new cable ladder was superior to hoisting methods commonly used in the U.S. at the time. While filming *Amazing Caves* in Greenland, the French were unimpressed with my custom-made climbing system and tried for two weeks to convert me. I pushed just as hard to convert them to mine. Equipment and techniques will continue to be perfected and debated by experienced cavers worldwide. I do not believe, however, that innovations in the foreseeable future will surpass spiders, the only true masters of the vertical climb.

Janot and Luc worked to dislodge the loose ice boulders near the rope, in hopes of avoiding accidental falls later. But more ice came loose. Even though the windchill above made the day bitter cold, the surface temperature had begun to rise above freezing, and everyone had to leave the cave. The filming and science would have to wait for colder temperatures in the morning. The next day, the Browns and Schultz at last recorded Nancy, Hazel, and Luc descending in the blue light of the moulin. Because of additional melting that occurred in the late afternoon, the platform had to be reset each time someone climbed to it. The filmmakers took advantage of this necessity by placing the platform at different heights, in order to film different sections of the drop.

At about one hundred feet below the surface, they were able to set up directly across from a small ice ledge where Hazel removed a section of ancient ice and quickly stored it in a sample tube. It was awkward work while wearing thick gloves, roped into a seat harness at least 20 stories above the pit's ground floor. But within the sample might be live tardigrades or new types of bacteria. She hustled out of the cave and rushed the sample to her liquid-nitrogen storage tank.

A few days after that, the film crew decided to try to lower the platform and the Mark II camera to a chamber near the bottom of the pit, 300 feet below the surface-deeper into an ice cave than a large-format camera had ever been. Schultz set the platform up about 25 feet above a small lake that had formed where water had backed up from the narrow channels far below. Mike Brown rappelled down to join him, carrying a heavy load of film. Just before he touched down, he noticed a large block of ice clinging to the wall nearby, with a nearly level surface. It seemed firmly wedged between two large vertical ribs of ice. Mike suggested over the radio that the small natural platform might provide a safer, less exposed camera position than the platform.

Janot rappelled down from the surface to examine the spot; he agreed that it looked more protected. For several days, Janot had worked to dislodge loose ice higher up in the pit, but here in the greater depths, 40 or 50 feet above the deep camera position, many loose, jagged boulders still protruded from the sides.

Mike, Gordon, and Schultz continued to ferry equipment to the ice block. On his second trip down, Schultz touched one wall gently with his ice ax and a huge crack shot across it like lightning for 70 feet or more. For a moment, he thought a single great sheet was about to peel off, sending tons of ice on Mike and Gordon below. But as quickly as it opened, the crack became still. Nothing more happened. The radios crackled with tension. Mike reported several "good sized chunks" of ice ricocheting at the bottom, pieces tumbling from elsewhere in the pit, but his flat block seemed to be well protected from the falls.

Finally, Gordon and Mike were ready to film the cavers and scientists working at the bottom of the moulin. But just as the Mark II began rolling, the ice block on which they had set up moved. It was a small shift, barely noticeable, but they were well above a lake of deep, freezing water, and neither was attached to a rope. Ever so slowly, stepping delicately while keeping their balance, like passengers leaving a car teetering at the edge of a cliff, Mike and Gordon anchored themselves to the wall. Once secured, they tested the floor, gingerly at first and then—in the manner of Wile E. Coyote testing a block above the Road Runner—more firmly. It didn't move again; they decided they could risk a couple of quick shots before abandoning the location. Janot rappelled to the cave bottom, passing the camera to touch down beside the lake in a beautiful, unbroken sequence.

The Browns suddenly realized that he had landed in a spot directly below their loose block, the sort of spot the coyote would have picked out for the Road Runner. They quickly shouted at him to move away,

One attribute some ice pits share with those formed in stone is a tendency for walls to bell out, away from a caver's rope, creating what is called a free rappel. When mountain climbers descend a cliff, bouncing against the rock face helps to slow them down. Controlling speed on an ice-slick rope, however, with the nearest wall far out of reach, requires expert handling of a descent bobbin, as Janot demonstrates here on his deepest descent of the expedition.

but the strange acoustics of the place combined with the language barrier made it nearly impossible for Janot to understand what they said. At last he looked up, and got the picture from the block itself. With a few deft, spiderlike moves, Janot angled away from below the block and found a firm anchor across the pit. But all present knew they had pushed their luck as far as they dared—there would be no samples gathered from the base of the pit today.

On the surface, Hazel watched the cavers emerge, their faces still white with tension from the experience below. She realized that the samples she had in hand would have to be enough: in two days the helicopter was due to remove her, Nancy, and most of the other Americans from the ice. The next day would be spent in the laborious process of derigging the pits and returning gear to camp.

Sometimes, the progress of science turns on improbable chance: Aristotle settles in his bath and develops a system for measuring volume. Alexander Fleming untidily leaves a bacterial culture sitting out, a mold spore drifts in, and penicillin results. Less celebrated are those times when chance and circumstance prevent science from happening. Due to a lost declaration form, the IMAX cameras, tons of gear, dozens of rolls of exposed film, and Hazel's liquid nitrogen tank sat for days in a hot warehouse near the Los Angeles Airport before being cleared by U.S. customs. The film and gear were undamaged by the heat. The tiny bits of ice and their ancient inhabitants, sealed inside the canisters' inner housing, were unfortunate victims: only a few ounces of terribly expensive drinking water remained inside the collection tubes when Hazel at last removed them in her Colorado lab.

After delivering an incisive, unprintable commentary in Bristol vernacular, she vowed that the next time she searched out strange new life, she would do it someplace warmer.

Despite warm weather that kept them from a world depth record, the team returns from the depths of Minnik II with a deeper understanding and appreciation of the massive forces at work beneath the ice and the minute creatures that hide there.

WATER

RIVERS BENEATH THE YUCATÁN

WE HAD UNDERTAKEN
OUR JOURNEY TO THIS PLACE IN
UTTER UNCERTAINTY AS TO
WHAT WE SHOULD MEET WITH;
IMPEDIMENTS AND DIFFICULTIES
HAD ACCUMULATED UPON US,
BUT ALREADY WE FELT INDEMNIFIED
FOR ALL OUR LABOR.
WE WERE AMID THE WILDEST
SCENERY WE HAD YET FOUND IN
YUCATÁN; AND, BESIDES THE DEEP
AND EXCITING INTEREST OF THE
RUINS THEMSELVES, WE HAD
AROUND US WHAT WE WANTED
AT ALL OTHER PLACES,
THE MAGNIFICENCE OF NATURE.

John L. Stephens
At The Ruins Of Tulum In 1842
Incidents Of Travel In Yucatan

Preceding pages: A diver negotiates a branch-
ing tunnel within the karst aquifer of the
Yucatán peninsula, site of the world's longest
known underwater cave system. Right:
Stalactites such as these form only in air-filled
chambers; most grew during the last ice age

CHAPTER 5

LAND OF CENOTES

F ine clay silt lined the bottom of the constriction. As Hazel squeezed through, her side-mounted air tanks rang like bells against the hugging limestone. She pushed forward. A single flip of her fin sent silt billowing out from below, a spreading brown cloud quickly engulfed her like a hungry organism. ■ Her light could penetrate no more than a few inches through the suspended clay. She was alone with her training, the techniques drilled into her that keep cave divers alive in situations exactly like this. The main things, she knew, were to keep calm and to maintain a gentle but steady grip on the dive line. Sight would be useless until the silt-out passed; only the stretched nylon cord loosely encircled by her fingers could return Hazel to visibility and ultimately to breathable air. ■ She followed the line out of the cloud, exactly as instructor Dan Lins had taught her to do six weeks before. Although she couldn't see him now, she knew Dan was pointing the rays from a brilliant searchlight that soon broke through the watery gloom. The film crew, some of the world's best cave divers, lingered only a few feet away, recording her passage through the squeeze on large-format film. Despite this knowledge, she felt her breathing rate increase, excitement speeding the depletion of her air. The squeeze and the silt-out had been carefully choreographed on the surface, and Hazel remained focused on her part. Still, in the back of her mind she knew that the tense scene she was

Above: John Lloyd Stephens procured a government posting to fulfill his dream of searching out rumored Maya cities. Right: Cave divers have modified gear in order to safely penetrate flooded chambers.

recording here, 50 feet beneath the jungle-covered limestone of Mexico's Yucatán Peninsula, matched what had been the final moment of life for many cave divers before her.

Cave diving is a sport that allows no second chances. Any error—equipment failure, rockfall, a moment of confusion—can be fatal. Fortunately, proper training and cave certification equips divers to overcome the most common dangers. Most of the half dozen or so people who die every year in underwater caves are open-water divers, unfamiliar with the special dangers of the cave until it is too late to do anything about them.

A submerged cave presents challenges completely foreign to even the most experienced ocean diver. Ceilings may collapse with little or no warning. Tight constrictions can snag hoses and tanks. A single diver's passing can stir up silt clouds, diminishing hundreds of feet of visibility to a few inches. Huge, irregular chambers can disorient divers to the point where they literally cannot tell up from down.

Perhaps the greatest dangers of all are those inside the diver's head—the opposite but equally deadly mental hazards of panic and overconfidence. Even seasoned military dive instructors have become disoriented and drowned after venturing only short distances into caves routinely explored by certified cave divers. Yet despite the sport's myriad dangers, deaths are surprisingly rare among those who obtain proper training and certification. The fatality rate of certified cave divers is much lower than that for sky divers or bungee jumpers, for example.

There are two well-defined levels of cave certification. At the first, "cavern diving," the diver is allowed to enter a cave, but cannot travel beyond a direct line of sight from the entrance, nor below certain depths. To become a cavern diver, one must master a number of basic safety drills unique to "overhead" environments—places where a diver cannot simply ascend to an airy surface if something goes wrong. A small percentage of cavern divers progress to the next level of certification, "cave diving," which requires such tests as laying a guideline and exiting a cave with no light, by feel only. Hazel had earned her cave certification specifically for this expedition, running through a variety of underwater drills that had simulated nearly every potential cave emergency. She felt confident that her instructor had imparted the skills she would need for microbial investigation of this watery world. Yet she remained constantly aware of the well-defined depth and distance limits to which certification extends. For a few divers at the cutting edge of exploration, those who choose to venture beyond the circumscribed ranges imposed by certification, the shadow of death lurks over every trip.

In 1997, Rob Parker, a celebrated British cave diver known for his work in Caribbean "blue holes"—the characteristic limestone pits in shallow seas around the

Mist rises from the swamp surrounding the entrance to "Jaguar Cave," so-named by members of the MacGillivray Freeman expedition because of fresh jaguar tracks they found in soft mud near the cave's freshwater stream.

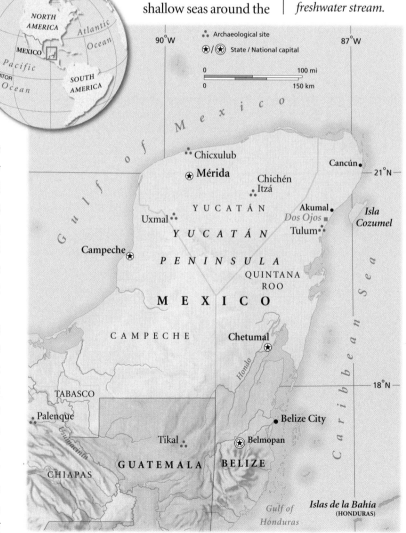

Bahamas—died at the age of 35 while returning from exploring a "bottomless" fissure he had entered at a depth of 220 feet. After a 20-minute survey of the fissure, Parker and his dive partner, Dan Malone, turned around, following the "rule of thirds," which requires divers to save two-thirds of their air supply for the trip out. As they worked their way toward the entrance shaft, Parker appeared to lose consciousness and drifted downward.

Malone nearly drowned himself—using up most of his air reserves and losing his fins-in the process of pulling Parker back from the abyss and reviving him. Parker signaled that he was okay, and the two men continued on, until they were forced to separate at a short constriction. Upon reaching the far side, Malone turned and waited, but after five minutes, nearly out of air, he was forced to surface. Parker never came out. Other cave divers who later analyzed the incident concluded that, despite over a thousand cave dives, many of them at greater depths, Parker had succumbed to nitrogen narcosis, "the rapture of the deep" that can lull divers into a deadly sleep. Other less experienced divers who have become "narced out" at depths over 200 feet have become lost within relatively straight passages, have failed to switch from empty air tanks to full, and have even abandoned their scuba gear in futile attempts to swim to the surface.

In 1994, the man that many considered the world's foremost authority on cave diving, Sheck Exley, died at the age of 45 while making a difficult attempt at a world scuba depth record nearly a thousand feet down a massive cenote—a sinkhole, or vertical shaft of water—in the Mexican state of Tamaulipas. Unlike the oceanic blue holes of the Bahamas, the cenotes of Mexico lie well inland, often surrounded by thick jungle. These remote portals include both the deepest and longest underwater caves ever explored. Exley was the author of leading textbooks on cave diving. He had personally trained many of those now regarded as top cave-diving instructors. He had set world depth records several times previously and held several world records for horizontal distance traveled in a cave. On his final attempt, while breathing from tanks of exotic mixed gases to be used at different depths in the massive pit, he simply dropped out of sight.

His partner in the record attempt, Jim Bowden, had first discovered the deep cenote and had trained for two years to make the dive with Exley. Bowden experienced equipment problems before reaching the bottom and began his ascent a hundred feet earlier than planned. He assumed Sheck had continued to the target depth and would soon ascend beside him. Sheck never returned. Other divers later speculated that he was the victim of an unusual condition caused by extreme pressure; but at the great depth to which he had descended, any number of things could have gone wrong. Anything wrong could have proved fatal.

As divers push below 600 feet, they are exposed to high-pressure nervous syndrome (HPNS), a neurological reaction to rapidly increasing pressure. The eyes shrink, causing divers to see flashing auras around people and objects. Accurately depicted in the science fiction film *The Abyss*, HPNS can cause violent body tremors, convulsions, hallucinations, and death. Naval and oil-company divers successfully battle the syndrome by descending slowly, inside submersible habitats—expensive steel capsules that would be impossible to transport into most caves.

Divers with COMEX, a French petroleum firm, have worked successfully below 2,400 feet. But such divers descend in fully equipped habitats over a period of days, acclimating to the depth while watching videos and eating TV dinners. After spending days at the bottom, conducting short dives between long rests, the slow rise to the surface may take weeks. Cave divers with no habitat to retreat to and only minutes of breathing time on their backs face a near certainty of HPNS from very deep descents. The syndrome often hits in combination with "compression arthralgia," a condition known to navy divers as "no joint juice" because it feels as if their knees, elbows, and wrists have suddenly rusted solid.

I was among those waiting for Sheck at the surface that day.

I had met him shortly after I first began caving in Florida in 1980, and had followed his cave-diving career over the years. I had joined his expedition with an assignment to write about his record dive for *Sports Illustrated*. Instead, I wound up writing his obituary. His body was recovered only by chance a few days later, snagged on one of the lines that had held spare tanks for his assent. I had always held a deep admiration for Sheck and three or four other legendary cave divers that I considered in his league. But I had personally resolved to limit my caving to places above the water table.

So I hesitated for a moment in the fall of 1999 when I was invited by MacGillivray Freeman Films to a join a cave-diving expedition to the Yucatán . Throughout the 1980s I had worked among surface support crews for cave divers in Mexico, Florida, West Virginia, upstate New York, and other locations, but I had not accompanied a diving trip since Sheck's fatal dive in 1994. I was encouraged to learn that the expedition's underwater film crew would be supervised by Wes Skiles, one of the most experienced cave divers around, the most experienced by far at photographing underwater caves. And I knew that, compared to the caves that claimed Parker and Exley, the hidden rivers of the Yucatán present relatively "friendly" environments because of their generally shallow depths. Even so, the task of moving a large crew deep into the cave was likely to be fraught with hazard.

Nancy vowed to cave only in air-filled passages even though the film company had offered, should she wish to join the underwater team, training in cave diving. "I want to stick to places where I can get lost for a few hours without running out of air or suffering an embolism," she said. It was a decision with which I could fully sympathize. Yet despite a healthy dose of fear, if not for myself then for those attempting to bring a large-format camera to such a difficult environment, I was fascinated by the scientific goal the film crew had set itself: the complex task of filming exploration of the halocline, a shimmering layer where fresh groundwater meets saltwater from the sea—a potential new environment for extreme microorganisms.

The year before, I had collected unusual organisms from Villa Luz, a sulfur-rich cave in the Mexican state of Tabasco, 500 miles west of the region to be targeted by the current expedition. I had shipped one of the samples to a scientist at NASA's Johnson Space Center, who had been so intrigued by it that he had visited the cave himself a few months later. The mineral-based microbial ecosystems of Mexican caves were providing living models of biology that might survive beneath the presumably sterile surface of Mars. If subsurface groundwater had existed at some time in the Martian past (or perhaps even remained today), some studies had suggested that it would probably be rich in mineral salts. In contrast, most of the water in caves is fresh, surface runoff purified by flowing through spongelike limestone pores. But in cave-rich coastal regions like the Yucatán, seawater can penetrate miles inland, creating a halocline at the boundary zone. Microbial life surviving in the halocline might be even more Mars-like than the strange organisms from Villa Luz.

I said yes and set to work checking my vertical harness and climbing gear; finding my fins, mask, and snorkel; and packing my dusty, well-traveled duffel. For reading on the two flights down, I dug out a paperback I had picked up in a Tampico bookstore five years earlier, *Incidents of Travel in Yucatán*, published as a two-volume set by John Lloyd Stephens in 1843. After the 1994 Zacatón expedition ended in tragedy, I had set the book aside unread. Now, as I opened it for the first time, I realized that I was bound for the heartland of the Maya in the company of an excellent guide.

In 1839, Stephens, a 33-year-old Wall Street lawyer, adventurer, and author was appointed American ambassador to the government of Central America. The only problem with the job was that he had no idea whether such a government existed. The United Provinces of Central America had declared independence from Mexico in 1823, but the nation had slowly foundered since its inception. By 1839 most of the states of Central America, along with the bordering states of southeastern Mexico, were essentially without rule. It was not a land particularly hospitable to travelers.

All the same, Stephens persuaded President Martin Van Buren, a fellow New Yorker, to authorize him to travel to Central America in order to seek out whatever seat of government might exist. His personal goal was to reach the Mexican state of Yucatán. At its founding, Central America had urged the southern Mexican states of Yucatán and Chiapas to join the fledgling union. Both had stayed with Mexico, but six years of unstable Mexican rule under Santa Anna (interrupted by several other short-lived presidencies) had distanced the ruling classes of both states, and both were reported to have considered seceding in favor of Central America. Stephens, ostensibly, was to investigate the political mood of Yucatán on his journey. Far beyond his political mission, however, Stephens dreamed of seeing ruined cities.

His partner in exploration was Frederick Catherwood, a British artist living in New York. Both men had already found audiences for their respective writing and painting of ruins visited in dangerous places, notably in Egypt and Arabia. They planned to make the first extensive scientific surveys and paintings of ruins rumored to hide in the steaming Yucatán jungles. Shortly after their arrival in what is now Belize in the fall of 1839, the two were able to travel to Honduras to explore the

Artist Frederick Catherwood's 1846 map shows many of the ruins and caves he and Stephens surveyed during their several expeditions. The pair braved poor and nonexistent roads, disease, bandits, and revolutions during their archeological treks.

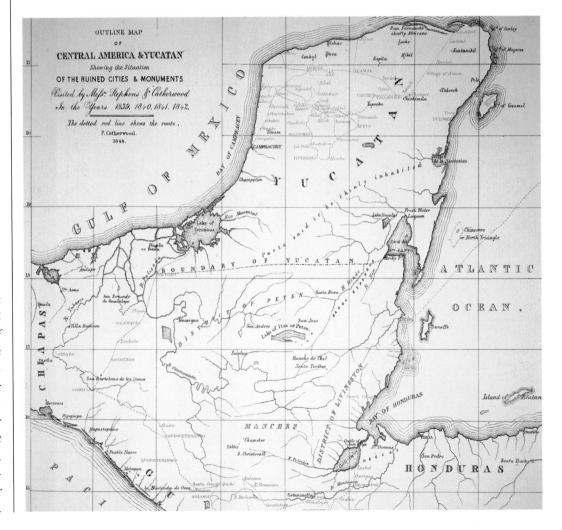

ruins of Copán, in which they recognized the hallmark of an advanced—and vanished—civilization. They measured and sketched the buildings and carefully copied the many hieroglyphs found on the great stone monuments, unaware that they were the ambassadors who would first bring the lost realm of the Maya to the English-speaking world.

Stephens purchased Copán for $50 from the alcalde of a nearby village. He arranged to have some of the mysterious carved stones shipped to New York to form the core of a new museum—the American Museum of Natural History, where the stones can still be seen today. He and Catherwood continued their journey through Guatemala, El Salvador, Nicaragua, and Costa Rica, encountering along the way armed bandits, torrential storms, black clouds of insects, impassable swamps and mountains, and devastating jungle fevers. They found additional ruins in several places, all bearing sim-

ilar mysterious symbols, but no traces of a Central American government. After months of travel, the two adventurers at last reached Yucatán, the land that had been their goal from the start, and there they found the ancient city of their dreams: Palenque.

By then it was June 1840, the height of the rainy season, with insects and disease taking their toll. A bite on Stephens's foot caused it to swell several times larger than his boot, and Catherwood was so weak from an intestinal flu that he could barely lift his pen to draw. Stephens was convinced that the hieroglyphs covering the buildings and monuments recorded the history of a lost civilization. Despite his illness, Catherwood made painstaking copies. Stephens's foot soon healed, but Catherwood's condition continued to deteriorate. Unable to continue their explorations, the two began the month-long journey to the Gulf of Mexico and home to New York. They vowed to return, having

Catherwood's portrait of Aké, near Merida shows a forest of stone columns arranged around a cenote. Stephens observed that the largest Yucatán monuments were typically built on or near cenotes, sacred in Maya religion. This particular site was the last the pair visited; it remains incompletely explored today.

received "vague, but, at the same time, reliable intelligence of the existence of numerous cities, desolate and in ruins, which induced us to believe that the country presented a greater field for antiquarian research and discoveries than any we had yet visited."

In 1841, Stephens's *Incidents of Travel in Central America, Chaipas and Yucatán*, profusely illustrated by Catherwood's drawings and paintings, became the first to argue that ruined cities to the south had not been built by ancient Egyptians or Phoenicians—theories then fashionable—but were in fact constructed by the same native Americans whose descendants now lived amid the ruins in simple villages. Stephens concluded that some of the cities had been inhabited and thriving at the time of first Spanish contact, fading into obscurity only after the native kings had fallen to the conquest. Later research would show that most of the ruins had been abandoned for two to three centuries when the Spanish arrived, but Stephens was the first author to give the Maya proper credit as

builders. Some authorities believe he was correct in the case of Tulum, which may still have been occupied at the time of first Spanish contact. They cite as evidence a mural found there depicting Chac, the rain god, riding a four-legged animal, a practice unknown until the arrival of Europeans.

The book was an instant success, and Stephens and Catherwood made plans to return to the Yucatán Peninsula in order to more thoroughly explore its ruins, as well as the natural caverns and cenotes that seemed associated with so many of the ancient sites. Their second journey took place between October 1841 and July 1842, and resulted in the two-volume set *Incidents of Travel in Yucatán*, published in 1843. Both books still provide a useful introduction to the region.

I drove southwest along and headed out of the city on Avenida Tulum, a street named after the ruined Maya coastal city that lay about 150 kilometers and five centuries from

Cancún. The white temples and castlelike buildings, castillos, of Tulum, first excavated by Stephens and Catherwood, are perched on a rocky cliff above the Caribbean Sea. The underwater caves to be targeted by the expedition lay hidden in nearby jungle.

The last time I had driven alongside a coastal forest en route to a tropical cave had been on the Big Island of Hawaii, where I had accompanied Larry Mallory, a microbiologist and medical researcher, into several lava tubes. By studying the unusual chemical properties of cave bugs there and elsewhere, Larry had isolated substances that showed amazing promise as new cancer-fighting drugs as well as antibiotics of a whole new class. But I knew that Larry would have no interest in the location I was headed to now: he limited his collection areas strictly to U.S. state and national parks, because bringing a new drug based on a living organism to market—from discovery to patent, through testing and FDA approval—required on average of ten years and 50 million dollars. Only the largest international drug companies could afford the process.

To count on their backing, Larry had to be able to prove the provenance of, and his permission to collect, every microorganism in his lab. He had learned that some foreign governments would lay claim to as much as 70 percent ownership of any patented product based on discoveries made on their soil. In order to avoid the threat of such claims, the major drug companies refused to back researchers who collected natural products from any but an approved handful of host countries. Mexico was not on the approved list. Luckily for scientists like Hazel Barton who were doing basic genetic research, the provenance of a newfound organism didn't matter, at least in any legal sense. Her goal was to answer broad questions: What is it? How does it survive? What else is it related to? The answers to such questions could guide other researchers like Larry in asking the narrower—and legally delicate—question: What use can we make of it?

As I fought the Cancún traffic, the city seemed to have more in common with Miami than with the Maya. The billboards were in English, and many of the restaurants were American chains. Each beachfront resort seemed several stories taller than the last, and the dress and demeanor of the crowds strolling the streets bespoke the American Midwest. It was hard to believe that 30 years ago there had been nothing here but a spit of sand and a few fishermen. Just as developer Carl Fisher had invented Miami Beach and ushered in a century of growth in Florida, the Mexican government had created Cancún from whole cloth, an instant tourist city now sprawling far beyond its borders. The coast south of Cancún was the fastest growing real estate market in Mexico. Those beachfront stretches not already claimed by huge, resorts bore billboards proclaiming resorts to come or, more rarely, screaming that the land was available to the highest bidder.

My rental car had not come with a full tank. As I gassed up at a convenience store in one of the southern suburbs, I heard at the next set of pumps a conversation in the dissonant glottal stops of the Mayan tongue. "There is no diversity of Indian languages in Yucatán," Stephens wrote. "The Maya is universal, and all the Spaniards speak it."

Five years ago, the road south of Cancún was little more than a potholed track; now it was broad and smooth. The highway swung inland, the beach resorts retreating beyond long private drives. I was able to accelerate to 100 kilometers an hour, warm tropical air whipping through the open windows. Twenty minutes outside of the city, I began at last to see traces of the Yucatán as Stephens must have seen it. To the right of the highway, away from the coast, lay an unbroken forest of vine-choked palms, oaks, and flowering tropical trees. Now and then a bare patch revealed the clean white limestone that built the Maya temples, which more recently provides the concrete for the temples of tourism. Further along, pictographic highway signs alerted me to roadside archaeological sites. I slowed to see low limestone ruins squatting amid small fenced-in clearings. Except for the slightly uneven nature of its hewn stones, one could have passed for the sort of square-block building that housed a municipal pumping station. Another was little more than a patio with a few steps leading nowhere. Yet as the sheer number of sites grew, I began to sense the fever of discovery Stephens must have felt.

I passed an enormous quarry, where large swaths of jungle and underlying bedrock had been removed as though lifted whole. Giant trucks rumbled past exposed cliffs at the edge of the quarry. The white cliffs were pocked by black shadows, entrances to the cave systems that underlay the entire peninsula. There are no significant rivers in the Yucatán, where all water flows underground. The frequent cenotes, vertical shafts to water, and the more traditional caves provide windows into the rivers below. Long before the Maya came to dominate the region, the first inhabitants learned to find water underground. Even in the longest droughts, some cenotes and dripping cave springs did not go dry. The caves and cenotes took on a religious significance to the Maya and their precursors to a degree almost unknown in other cultures. Caves were sacred places, sites of ritual, initiation, and sacrifice, and the locus of great myths; some of the greatest artifacts of Maya art and culture had been found in them. Nearly every great ruin on the surface had grown above a natural or artificial well, symbolizing a mystic connection to the mythic underworld.

I turned off the main highway at Akumal, the expedition's base. Unlike the clustered tents and cot-filled barns I'd experienced on other trips, Club Akumal offered a tropical paradise. Comfortable, palm-shaded cottages—complete with air-conditioning and Catherwood prints on the walls—sat only a few steps from a beach of sand as bright and fine as confectioner's sugar. Coral reefs shimmered from beneath the clear blue waters of a protected lagoon. A short walk from my cottage, I found a bar and grill beneath a

Catherwood's painting depicts the precarious "ladder" used by natives of the village of Bolenchen to retrieve water from a natural pit over 200 feet deep. In a region where all rivers flow underground, caving skills were often essential for survival, leading caves to play a central role in Maya mythology.

thatch-roofed palapa—the ubiquitous wall-less shack of the Caribbean. Lunching on shrimp cocktail and cerveza with lime, I divided my gaze between white breakers hitting the reef and CNN on the overhead TV. Stephens and Catherwood had definitely endured lesser accommodations.

The expedition consisted of two independent teams: a dive crew to film the underwater work and a topside crew to film the and air-filled caves. Both crews were out in the field, I had learned upon checking in, not expected back before dark. Since I had no hope of finding their remote locations on my own, I decided to use the remainder of my first day in the Yucatán to visit Tulum.

When Stephens and Catherwood visited the ruined cities of their second journey, they hired crews with machetes to chop away the growth that obscured most of the buildings. The clearing had continued more or less unabated in the 14 decades since. In the 1930s, and again in the 1950s and 70s, collapsed arches, walls, and buildings at many sites had been restored by architects and archaeologists working in concert to provide tourists with a more "realistic" experience of the Maya world. And at sites where the main structures remained relatively intact, such as Tulum, buried monuments and shrines missed by the first explorers had been excavated, the surrounding grounds leveled and landscaped.

The buildings of Tulum are less spectacular than those of Palenque or Chichén Itzá, but the setting—on a verdant hill above a crashing turquoise sea—looks like something imagined by Maxfield Parrish. The view, combined with the proximity to Cancún, means that busloads of tourists arrive daily. Mexican, American, and European visitors swarmed over the site as I walked through the narrow stone gate in the western wall. Dozens of large and small structures stood in a grassy, parklike field. People posed, or waited in line to pose, for photographs in front of virtually every one of them.

The name Tulum comes from the Maya word for "dawn" or "rebirth." The site's

Tulum is unique among Maya cities for its coastal location, atop a limestone bluff. Legend had it that the top of the central "castle" was the first point in the Maya world to be struck each day by rays of the rising sun. Some evidence suggests that the walled city was still occupied by the Maya when Spanish explorers sailed past in 1518.

highest structure, called El Castillo or the Castle, perches at the peak of the cliff rising above the sea at the eastern edge. It would have been the first point touched by dawn in the Maya world. I knew from the shooting schedule I carried that Nancy had visited the ruins with the topside crew the day before. The film crew had arranged to come early in the morning, before the site was opened to the public, to record the dawn light that had made the city so imposing in Maya times.

Stephens was the first explorer to clear and map Tulum's walls. A limestone barrier up to 20 feet thick encloses the rectangular com-

Some underwater constrictions are so tight that cavers must remove their tanks and shove them through ahead by hand. Squeezes can cause not only claustrophobia, but increase silt-outs and the danger of collapse, as divers must make close physical contact with sediments and rock walls.

pound on the northern, western, and southern boundaries; the steep sea cliff takes the place of an eastern wall. Archaeologists believe Tulum was first built during the Early Post-Classic period, from AD 900-1200, or roughly the same period in which Vikings settled Greenland and Vinland. The Maya city-states were frequently at war with one another during this period, until the fair-haired, bearded god-king Quetzalcoatl ("the feathered serpent") arrived from Tula in the west and united the empire. The walls of Tulum made it as impregnable as any castle of medieval Europe.

I walked along the cliff, passing young couples gazing at the crashing spray, children scampering over ancient altars. Although the waves

of countless hurricanes had carved niches and alcoves into the cliff face, I saw no true cave entrances. I hiked north from El Castillo and through a small ruin that a sign identified as "The House of the Cenote" because of a small well at its center. Working my way down to the back of the building, I found myself alone in a small depression—a natural sinkhole.

The steep walls muffled the crashing waves and shouting children only a few yards away. No one saw me approach the small, brackish pool at the bottom. Waterbugs danced over the surface. Beyond the shallow pond stood a narrow slot, its sides slick with mosses and a yellowish scum I suspected was bacterial. I pulled a small flashlight from my pocket— always handy when hiking over karst, the scientific name given to limestone landscapes punctuated by caves, springs, and vanishing streams—and skirted the pool to reach the small entrance. It curved a few feet to the right and ended in a tiny alcove, much too short to call a cave but bearing the remains of ancient stalactites in its ceiling. Without this vestigial cave and its small spring, I suspected, the city might not have grown here at all.

As I was poking through the shadows and depressions of Tulum, Hazel was struggling through a spot that no tourist would ever visit, deep into Dos Ojos. She followed the darting rays of Dan Lins's movie lamp. From the beyond the gloom of the silt-out, a strangely metallic voice came through the water. "Great, now wiggle your dive light back and forth as you come out."

Hazel complied. The disembodied voice belonged to Wes Skiles, who had worn a full face mask with a built-in microphone and underwater PA system in order to direct the complex scene. As he directed, underwater camera operator Howard Hall captured Hazel bursting forth from the silt cloud.

"Great," Wes said. "Now catch your breath and let's set up in the next chamber."

They swam forward, each diver holding a piece of equipment with one hand and tracing the dive line with the other.

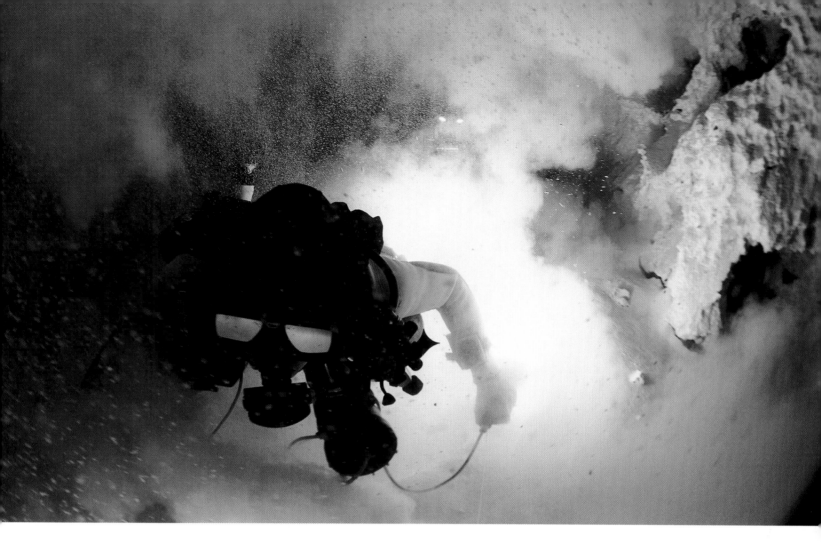

I climbed out of the sink, wiping the slime off my hands on the mown grass, and began following the north wall. A short distance from the House of the Cenote was a narrow arch, smaller than the western gate through which tourists entered Tulum. There was no barrier across it as I had seen at other openings in the ruins, so I walked through. From El Castillo, I had noticed a large beach house situated a hundred yards or more north of Tulum, but amazingly, the land between the ruin and the house appeared undeveloped and pristine.

Stepping through wall was like stepping into one of Catherwood's paintings. Thick roots hugged the stacked stones, in some cases spreading them apart as though trying to undo the restoration of recent decades. Ahead, snakelike vines twined around the smooth, cinnamon-colored trunks of flowering trees. Clouds of insects, strangely absent from the manicured grounds, hovered about my head as though tourists were fair game on this side of the wall.

I recalled what Stephens had said in deciding not to search for ruins beyond the walls: "A legion of fierce usurpers, already in possession, were determined to drive us out, and after hard work by day, we had no rest at night:

> There was never yet philosopher
> That could endure toothache patiently.

And I will venture to say that a philosopher would find the mosquitoes of Tulum worse than any toothache."

The overgrown lot in which I stood occupied only a few acres. Still, it felt like the sort of jungle where new discoveries could be made, the sort of jungle where perhaps the team I was about to join had found something this very day. The sun slanted through the vines, dipping toward the west. I savored the quietly fierce buzz of insects a moment longer, then retraced my path back through the wall to the tourist attraction, following the retreating crowds to the parking lot. The two teams should be returning to Akumal soon, I thought. It was time I met the members of the expedition. According to the schedule I carried, we had a 5:30 call the next morning.

Shrouded by silt, Hazel emerges from a tight underwater squeeze in the Dos Ojos system. Because of the isolation imposed on cave divers by tight passages, each person carries a redundant air supply and spare regulator, similar to the back-ups built into NASA space suits.

THE CAVE DIVERS

Wes Skiles paced the crowded, brightly lit wooden dock just inside one of the twin entrances to Dos Ojos—the "two eyes" of one of the world's longest underwater cave systems. The southern accent of the balding, solidly built, 40-something cave diver echoed from the concave limestone overhead. ■ "The level of diving is up an order of magnitude today," he said. "It's much more dangerous than yesterday because of distance. Beware of low air. Beware of route finding. There's a lot of clay silt on the floor." ■ Water dripped from stalactites, making concentric rings in the clear blue pool that surrounded the dock, occasionally sprinkling Skiles and his rapt audience. The ceiling stood 12 to 15 feet overhead, sloping until it met water in the darkness far beyond the dock. The rock-filled sinkhole outside the cave was perhaps a hundred feet in diameter. The dock sat in one of two large air-filled chambers extending from the lower edges of the sink. Wooden stairs led up behind the group to gray daylight, where a loud tropical downpour raged. Mixed with the sound of rain was the constant rumble of gasoline generators placed under a plastic tarp at the base of the sinkhole. These powered both the lights and an equally loud air compressor used to fill dive tanks. ■ A dozen divers, some in full wet suits, others in shorts and T-shirts, sat or

Above: Wes Skiles co-directed the expedition's underwater film sequences. Right: Howard Hall built the specialized housing that allows the IMAX camera to venture deep below the surface.

stood around Skiles as he paced and talked. They had wedged themselves among tanks and harnesses; racks of batteries; regulators; film canisters; a table of dive computers, radios, and other electronics; helmets; dive lamps; movie lamps; nylon mesh bags of fins and personal gear; underwater scooters; huge coils of power cable; an empty aquarium; and what looked like a miniature blue submarine hanging from a steel chain attached to heavy bolts in the cave roof. The scene resembled the military briefing of an elite underwater assault team. In a sense, it was.

"I want to see strong, strong, strong concentration," Skiles said.

"Y'all are here because you're cavers. So let's be gentle on the cave." He pointed to a map on a large white board, sketched in Magic Marker that morning by Hazel. "Beware of soda straws overhead when you go through here. Finesse your BC lifts. Keep your side mounts equalized." I knew that it was often difficult to avoid damaging delicate soda straw formations when crawling through a dry cave. Until now, I had never considered the difficulty of doing so when a buoyancy compensator vest was lifting you into the ceiling. This was one reason Wes had assembled such a seasoned crew—safe diving practices were so instinctive to them that they could devote more conscious thought and effort to protecting the cave than less skilled divers might manage.

The assembled team could have filled a Who's Who of cave diving and underwater filmmaking. Many were expatriate Americans living in the Yucatán, drawn south years earlier by the low cost of living and the unlimited miles of virgin cave waiting to be found. Others had driven from as far as California and Florida. Nearly all were certified cave-diving instructors; nearly all had participated in the body recoveries all too common to the deadly sport.

Wes's chief collaborator for the underwater sequences was Howard Hall, who had more experience filming underwater with large-format cameras than anyone alive, just as Wes had the most experience filming submerged caves. As Skiles talked, the often taciturn Hall gently

Some cenotes, such as this jungle well—its exact name and location a secret closely guarded by cave divers—are undercut on all sides from previous collapses. In order to enter this window into the aquifer, divers must descend 75 feet on free-hanging ropes or cable ladders, then later haul their gear (and themselves) back to the surface. The overgrown cenote walls often hide snakes, poisonous plants, and bees and hornets.

Ensconced in a bubble of light, divers approach the halocline, a fog-like layer separating freshwater and saltwater that has penetrated inland through the aquifer from the nearby Caribbean. Because the halocline represents a mixing zone between differing water chemistries, microbiologists like Hazel Barton regard it as a likely collection area for unknown "bugs."

cleaned the lenses and dials of the underwater camera housing he had invented. Hanging from its chain, the bright blue behemoth weighed over 300 pounds; underwater it would be essentially weightless. Working under a National Geographic Society grant, Hall had spent two years and $90,000 building the housing. He had steered it over coral reefs and wrecks, through a frenzy of feeding sharks, but never before had he brought it to an environment as dangerous as an underwater cave.

All eyes and ears at the Dos Ojos dock remained fixed on Wes as he went over the procedure for lighting divers who would pass by on electric scooters. I was reminded of the briefing that began each episode of the old *Hill Street Blues* series. I kept waiting to hear Wes say, "Let's be careful out there." Instead he simply said, "I guess that's it. Let's go."

One by one, the divers eased on the harnesses that held their tanks. Some used typical back mounts, while Hazel and two others wore side-mounted tanks—better than back mounts for passing constrictions. All stepped onto the submerged wooden platform at the edge of the dock where the water was about eight feet deep. Director Steve Judson began discussing shots on the day's list with Wes. Upon meeting the topside and underwater filming teams and hearing them discuss plans the night before, I had been struck by the differences between filming for IMAX theaters and filming documentaries for television. Everything is bigger for an IMAX scene: the camera; the lighting requirements; and the size, weight, and (especially) the cost of the film stock. Thus crews are of necessity larger, the process more akin to that of a big-budget feature film than a traditional documentary. Even in an "ordinary" outdoor environment, an IMAX shot has to be carefully planned to succeed. In an environment as difficult as a cave, especially an underwater cave, the planning must be very detailed indeed.

The amount of gear and the strictness of routine required just to survive a cave dive are daunting. Add to those basic requirements of material and drill a camera in its massive hous-

ing, hundreds of pounds of lighting equipment, thousands of feet of cabling, and the world's heaviest and most expensive film—and imagine using it all in a professional manner while paying constant attention to the requirements of the cave dive—and you begin to get an inkling of the challenge the underwater crew faced. To have faced that challenge with an unplanned dive, depending on chance alone to present both photogenic views and reasonably safe conditions, would have been all but impossible.

So while the project presented ample reality—real caves, with some new exploration planned and actual scientific samples to be gathered, and real danger as well—it was nonetheless a scripted reality. Hazel and Nancy were cavers, not actors. Every movement Nancy made on rope or Hazel made behind an underwater scooter would be planned well in advance. For some of the most complex underwater scenes, a more experienced cave diver would double for Hazel (although the double's primary purpose on the team was to free Hazel for filming topside without interrupting the laborious and intricate shooting of underwater segments).

Like Nancy and Hazel, I sometimes found myself grinning at the Alice-in-Wonderland oddness of spending four hours with a dozen technicians in order to create a well-lit 60-second "moment." And yet there was no denying that during those 60 seconds, when they finally came off, we could feel the grandness of the art being made around us, could sense that maybe, perhaps, it really would be possible to give noncaving audiences the genuine feel of the underground realm.

Wes handed Steve a plastic underwater slate. Using a waterproof marker, Steve began writing in large block letters as he spoke. "Okay, four shots on the roll. Number 1 is the scooter return, coming back after the silt-out. On Number 2, Hazel should be in front, laying line. Number 3 should have Hazel in back, not laying line. And Number 4 is the scooter crossover."

"Got it." Wes clipped the slate to a D-ring of his dive harness.

The lighting instruments were already in the

DINOSAURS, METEORS, AND CAVES

A SHORT HISTORY OF A LONG TIME

BY CHARLES E. SHAW, PH.D.

Sixty-five million years ago, the Cretaceous Period came to a violent close when a meteor the size of Manhattan, traveling at more than 20 miles a second, slammed into a shallow seafloor due south of North America on what today is the north coast of the Yucatán Peninsula of Mexico. The present-day Maya village of Chicxulub stands at the impact center.

Upon entering the atmosphere, the leading edge of the meteor, ablaze from atmospheric friction, created a sonic boom that would have been heard around the world. A second or so later, the meteor plowed two-and-a-half miles into the solid Earth. More than four thousand cubic miles of limestone and other rock material was ejected from the crater or vaporized by the energy of the impact. At a distance of 55 miles from the impact center, the floor of the ocean collapsed downward more than two miles along a circular ring of faults that reached deep into the Earth's mantle. The resulting crater was more than one hundred miles across, with walls over a mile high. As shock waves moved outward through the solid earth, additional faulting took place at a distance of 80 miles from the center, forming a second ring around the point of impact. By the time it was over, a matter of a few minutes, a crater 160 miles across had been formed.

In seconds this cosmic accident erased a global ecosystem 183 million years old and drastically altered the subsequent course of Earth history. Fifty percent of all species then living were destroyed in this second largest of all extinctions. The demise of the dominant land animals of the Cretaceous and the ecological niches they left vacant set the stage for the evolution of large mammals, including humans.

The post-impact geologic history of the Yucatán Peninsula was strongly influenced by the size, shape, and depth of the Chicxulub crater. When the debris settled on the first day of the Tertiary Period, a chain of islands protruded above the sea over the outer ring, where the day before there had been a flat ocean floor. The islands were formed from a thick pile of debris that partly filled the crater and spread hundreds of miles in all directions. Even as they were being formed, the islands were eroded by huge tsunami waves set off by the impact, which sloshed back and forth across the Gulf of Mexico.

During the 65 million years since the impact, the peninsula slowly took on its modern aspect. Over time, the debris in the inner bowl of the crater was buried beneath 2,600 feet of Tertiary limestones. Half that thickness covered the crater rim and exterior ring. As the Tertiary limestones above the buried crater were compressed and compacted, a ghost of the crater's form appeared in the modern land surface. A low north-trending ridge curves across the east-central portion of the peninsula to the Gulf of Mexico and reflects the buried outer ring of the crater, like a dinner plate under a rug.

Farther west, a ring of water-filled sink holes, called cenotes by the Maya, have been dissolved in the soft limestone fill above the buried crater rim at 55 miles from the center, possibly following fractures caused by unequal compaction of the Tertiary limestones above the rim.

The cenote ring today is a major drainage path to the ocean for groundwater entering it from both the east and the west. Owing to its soluble limestone bedrock, the northern Yucatán Peninsula has no surface streams. Instead, a complex network of underground caves and channels has been dissolved to form a regional drainage system that carries water to the sea. The cenotes are collapsed portions of this underground network.

Extensive cave systems also carry water eastward to the Caribbean. These caves discharge groundwater through undersea springs and coastal lagoons all along the Caribbean coast. Westward flow of groundwater from the interior of the peninsula to the cenote ring and eastward flow to the Caribbean require that at some place between the two, the direction of groundwater flow must separate, east and west. This apparently happens near the surface expression of the outer ring, which also marks the modern surface divide of the peninsula. Thus, it appears that the shape of the crater that ended the Cretaceous Period controls the flow of modern groundwater 65 million years later.

chamber chosen for today's shoot, placed there the night before. In order to power them, a small hole had been drilled from the surface weeks before, with hundreds of feet of electric cable fed into it. Three crew members had lugged a generator through the jungle to this hole, located about a quarter-mile from the cave entrance. They would be signaled by radio to start the generator as the divers departed for their location. The night before, Steve Judson had told me that this film presented by far the most difficult locations he had ever encountered for any film, including Mount Everest. He added that mountaineer and filmmaker David Breashears, who famously carried the Mark II camera up Everest in 1996, agreed.

"Each location here in the Yucatán carries its own challenges. The rigging is tricky. You can never get the size crew you'd like into any passage, so you have small crews doubling up on several jobs. If we didn't have the combined talents of Wes and Howard, and all their years of experience, we'd have no hope of getting footage from the halocline."

Wes and Steve talked until interrupted by the squawking of Steve's radio. He spoke into it for a moment then turned back to Wes. "I guess that's everything. Have a safe dive."

"Roger that," Wes agreed, grinning.

Steve headed back up into the rain. Down a path several hundred yards away, the topside crew was shooting Nancy running from a cenote reconnaissance toward a dry camp.

The focus and intensity of the underwater team increased as they prepared to depart. As I tried to stand out of the way, I was surprised to hear my name shouted in a British accent.

"Yo! Mike! Could you step over here a sec?" Hazel Barton sat on the dock, wearing a lime wetsuit and a complex web of gear.

I walked over and she spun her back toward me. "Could you hook the clip hanging off the bottom of this tank to the D-ring on me back? I can't seem to reach it."

I knelt down, giving the tank a tug to clip it in. "There you go," I said. "Happy to oblige."

"Thanks." She turned to face me. "Listen, I think this roll we're about to shoot is all

they want me for today. What would you say to a little cavern dive around the sink later? There're some really beautiful formations in here."

I paused. "Well, I had planned on at least snorkeling around the perimeter," I admitted. "But I'm not cavern certified."

"But you're an open-water diver, right?"

"Yes, and I did a couple of very easy cavern dives once, long ago, with an instructor I was writing about, but still . . ."

"Oh, you'll be fine. We'll stay well in sight of daylight. We'll talk to Dan and Gary," she said, referring to Dan Lins and Gary Walten, two highly respected cave-diving instructors who were part of the underwater crew. "You know, there's a kind of commercial trip they do in the Yucatán all the time; it's called a cenote dive. You go with a guide and it doesn't require cavern certification. It's legit—they take tourists all the time. If Dan and Gary say it's okay, you've got to do it."

Even from the dock, I could see what appeared to be massive calcite columns shimmering in the depths. I had often heard other dry cavers describe the joy of "flying" through a well-decorated underwater cave.

"Come on," she said. "I expect you back here at 4 o'clock."

I gave in. "Sure. If it's okay with Gary and Dan, I really would like to see just a hint of what you guys are up to."

If instructors of their caliber said I could handle it, I was sure I could. If they advised against it, I'd have an excuse for backing out.

"Great." Hazel said. "Catch you then." With that she plopped underwater.

Two divers gently lowered the camera to Howard, who stood in waste-deep water to receive it. One by one the divers submerged, their lights and bubbles vanishing beyond a shelf of rock. Within ten minutes, I stood alone on the dock, looking out into the cave.

I turned and headed up the slick stairs into rain, hiking over rough terrain toward the topside crew, trying hard not to think about any of a hundred cave diving accident reports I had read over the years.

CHAPTER 7

RIVERS BENEATH THE BUSH

Unlike most underwater caves, the Yucatán systems were dry during the Pleistocene Ice Age, when they developed beautiful calcite mineral formations to rival those of the world's finest tour caverns. About 10,000 years ago, the caves refilled with water as the sea level rose from melting ice. Today golden calcite draperies, crystalline soda straws, giant columns, stalactites and stalagmites of every possible description hang in eerie silence. Several times over the next few weeks, the underwater team under Skiles and Hall carried the first light to ever shine upon these hidden realms. Even Skiles, who had filmed Yucatán caves for decades, later admitted he didn't truly understand just how beautiful the rooms were going to be until they were lit for the first time with the powerful HMI lamps the team carried. ■ With the lights on he'd find himself hanging suspended 20 feet above the floor, gazing at sparkling crystalline formations all around. He felt as though he floated inside an immense jewel. As the team waited to shoot sequences, it was easy to drift off into the dreamscape of the spectacular rooms and corridors. More than once, the temptation to wander off and stare at the surrounding marvels was dangerously strong. Skiles continually reminded the team to remember exactly where they were, and exactly what their job was, careful among so many hard-core explorers that none get lulled into the lure of the labyrinth. ■ Today they planned to spend as much as eight

Above: Brad Ohlund ascends an aluminum ladder hung above a cenote ledge chosen as a filming platform. Right: Drowned stalagmites of a previous geologic age stand like silent sentinels en route to the halocline.

hours underwater setting up and shooting sequences. One of Wes's goals as director of photography was to expose each image on film using sources of light that viewers would never notice or question. To accomplish this, his team had laid over 1,500 feet of cable, spidering out to five different light heads at varying depths and locations. When they had first scouted locations in Dos Ojos, he chose the most beautiful rooms regardless of their proximity to respective entrances. To deliver surfaced powered light to these rooms required a Herculean rigging effort by a team of cave divers and local Maya. Underwater, cave divers had to map their way into the rooms chosen for filming. Above, the Maya crew hacked through thick jungle to establish a base directly atop a chosen room.

Because the caves were part of an unconfined water table system, they were able to drill a cable hole into the chambers without harming the environment. With a mobile field camp built next to the cable hole, they were able to feed power from generators down into the rooms. The team's first effort at feeding cables into a chamber called Gothic Chamber proved fairly daunting. But the second attempt to cable a room over a thousand feet into the earth had almost ended in disaster.

Inadvertently, the surface team thought they had received the "feed cable" signal from the team beneath. Not knowing what they were actually feeling was the tugging weight of the cable itself, the surface crew accidentally let four 300-foot coils slide down the hole. Underneath, inside a billowing silt cloud, divers experienced a nightmare like a scene from the movie *The Abyss*: 1,200 feet of black, snakelike cable came spilling onto the cave floor, nearly trapping one of them. Fortunately, no one was harmed, and quick action by Gary managed to untangle the Gordian cable knot.

Before this expedition, no one had ever successfully captured on film the shimmering coalescence of saltwater and brackish water at the halocline, the mixing zone of freshwater flowing seaward and saltwater seeping inland from the coast. Because subterranean microbes depend upon the chemical conversion of minerals—as opposed to sunlight—for all of the energy in their food chain, the richest areas for new bugs tend to be natural interfaces between two types of chemistry. In such a mixing zone, the range of raw materials from which microbes can convert energy is far greater. For example, Tullis Onstott, a geomicrobiologist at Princeton, has found

thriving bacterial colonies at the interface between gold and quartz nearly two miles deep in a South African mine—yet the ore only a few inches away from the interface had been nearly devoid of life.

Here in the Yucatán, Hazel hoped that the differing chemistries as well as the differing fauna between freshwater and saltwater would present a rich area for new life. Michael Garman, an environmental biologist and cave diver in Florida, had recently observed thick bacterial mats at the halocline of a freshwater spring that emptied into the Gulf of Mexico north of Tampa. Haloclines occur only in stable, protected environments where a large volume of freshwater meets seawater. Thus the best place to find them is in coastal karst regions of the tropics, where heavy rainfall provides sufficient fresh groundwater to flow through the seaside caves. But a trained microbiologist had yet to study a natural halocline in situ—in the Gulf of Mexico or anywhere else.

When the divers first saw the halocline, it appeared as a mirage or an illusion. At times Skiles could almost swear he was looking at a cave filled with half air and half water. It was like swimming along the interface of oil and vinegar inside a bottle of salad dressing. Making filming even more difficult, once the halocline was disturbed, the two waters would mix, causing a blur-out in which divers lost all vision.

Working with special lighting techniques and camera angles, the team eased into the halocline chamber and prepared to film. Wes had come with a plan to use trash-can lids as "flags"—temporary covers for the movie lights. As the on-camera divers approached, underwater gaffers on the crew would uncover each light in sequence, making it appear as though the scene were lit entirely by the divers' flashlights.

Hazel carried with her several sterile plastic centrifuge bottles, which she would use to sample whatever life might hide within the shimmering layer. Just as it had never been filmed, the Yucatán halocline had never been

Buddy Quattlebaum steers an eponymous Buddymobile over a particularly rough bump while ferrying team members down one of the less-frequented jungle tracks he travels as an ecotourism guide.

Hazel collects a sample from the elusive halocline while local dive guide Jorge Gonzoles looks on. Biologists working in Florida caves have found heavy concentrations of microbes at haloclines— where fresh and saltwater meet—as well as at chemoclines, mixing zones where the freshwater aquifer intersects pockets of sulfur-saturated, oxygen depleted water. Until this expedition, no halocline in the Yucatán had been sampled; Hazel hopes to collect samples from Yucatán chemoclines in the future.

checked for microbial communities that might capitalize on the chemical reactions possible in only a zone where waters of differing salinities mixed. Using hand signals, Hazel and the dive crew signaled Wes that they were set for a scientific and cinematic first.

His robotic amplified voice penetrated the water: "Action!"

As he stood smoking topside, Manuel Noh Kuyoc looked as though he could have been any age between 40 and 65. He wore ancient cowboy boots with a light blue work shirt and tattered rayon slacks held up by a belt with extra holes punched in it to accommodate his narrow waist. His sparse mustache bristled beneath a prominent Maya nose. He stood just outside one of two doors to his traditional forest home, which at the moment was being made over as Hazel's "science hut" for the film.

The oval structure was built of wooden poles lashed together. It stood in a rocky clearing shaded by a half dozen fruit trees and surrounded by several other huts and outbuildings. In a small hollow ten feet from the cottage I saw the sort of rubbish that would have been indicative of a midden burrow, or refuse heap, found at an archaeological dig. Among the white limestone cobbles lay broken pottery, burned-out gourds, short lengths of native twine lashed around bits of wood, lemon rinds, coconut husks, a single plier, soda and deodorant cans, beer and Bacardi bottles, a rusted device that may have been some sort of insect sprayer, and a few empty tins of Tulip brand pork luncheon meat.

A swaybacked horse hobbled nearby tugged tiredly at a clump of grass, ignoring the heavy drops of rain. Manuel watched the horse for a few minutes then ground his cigarette butt under the toe of his boot and turned back to contemplate the activity in the house. The thatched roof seemed to be doing an excellent job of keeping it dry; the darkened interior looked warm and inviting. Pots, pans, machetes, more lanterns, garden tools, bags of soft drinks, and items of clothing hung

from wooden pegs ranged around the perimeter. Three bright hammocks slung on support poles had been pulled aside to make room for a rough wooden table bearing a microscope, a microbial staining kit, a kerosene lantern, a hot plate, and a pale blue English tea kettle.

Tom Cowan, director of the topside unit, removed the soft drinks from the area about to be photographed while Brad Ohlund, who had supervised much of the camera work on Greenland shoots, helped crew member Chris Blum set several lamps on tripods until their orange glow blended with that of the lantern on the science desk. Just outside the door directly across the hut from Manuel, Jack Tankard, and other crew members bolted together 20 feet of steel rails for a complicated dolly shot that would track Nancy as she ducked out of the weather into the hut.

I approached Manuel in the doorway, catching his bemused expression as Tom examined a bench behind the table, pulling from it a fairly new looking transistor radio, a solar-powered fluorescent lamp, and some candy wrappers. Manuel saw me taking notes. Although his Spanish was heavily accented and mine all but nonexistent, I understood when Manuel began to explain to me that he lived in a primitive hut without color television and indoor plumbing not because he did not know any better, but because he had chosen to do so.

For the past 14 years he had lived with his wife and children on a homestead named Tak Be Ha after one of several cenotes on the undeveloped land. The Kuyoks shared a swath of more than a thousand acres of virgin land west of the coastal highway with several other Maya families and an expatriate American who ran the business that made it possible for them to attempt life in the old way—who now approached from the direction of Dos Ojos, wearing nothing but ragged cutoffs and carrying a large walkie-talkie.

Few cave divers have heard the name Gordon Quattlebaum, whom virtually everyone who dives in the Yucatán knows as Buddy, the proprietor of the Dos Ojos Dive Shop. The tall American was bearded, with a dirty tan that reached from a balding head down over a string-bean body knotted with muscle. He seldom wore a shirt and never wore shoes. For the past decade he had led commercial cavern and cave dives as well as the complex exploration of the Dos Ojos system; he created the "guided cenote dive" of which Hazel had spoken.

Although he and the Maya who owned the land made their living off of tourism, Buddy was well respected by the region's environmentalists, and equally despised by a few of its developers because of his tireless efforts to protect the coast's indigenous sea turtles along with the rights and traditions of its indigenous peoples. He had cleared very little land at Tak Be Ha. The few roads he had built were nothing more than jungle tracks, over which he carried customers in a fleet of homemade vehicles called Buddymobiles. I had seen one of them at the cenote parking area, two miles down the muddy, potholed "main" road from the coast highway. It was little more than the engine and chassis of a large truck, with an open seat for the driver; a steering wheel; wood-block brake, gas, and clutch pedals; and a braced wooden bed for passengers.

Buddy spoke for a moment to Tom, director of the topside scenes, too softly for me to hear.

"When they come out, ask if we can have Hazel for an hour," I heard Tom answer. Buddy nodded, already headed back toward the distant noise of the generators, his bare feet gliding effortlessly over blades of limestone that poked up everywhere.

I walked around the hut, seeing it as the camera would see it. The scene was authenticity itself. I could not help but grin at how truly primitive every detail appeared. It would look very cool blown up on a huge screen. The movement of the camera along its rails, beneath dripping tropical branches, would need only distant drums in the soundtrack to send the film straight into Indiana Jones territory.

After more than an hour of setting up at the hut, Tom was ready to begin filming. Yet just as the camera began rolling—literally, down its steel track—the sky once more opened up. Tropical raindrops heavy as pebbles pounded down. Jack quickly unbolted the Mark II from its tripod and ran it into the back of a panel truck that had been coaxed down the bad road to the hut location. The rest of the crew scrambled to cover equipment in plastic. I hopped into the back of the truck with several others, crowding around in the open rear doors, sitting on cases or standing hunched in the back. The filmmakers decided to take advantage of the protected view of a jungle downpour by sending Nancy out into it, reshooting the sequence in which she approaches the hut.

"This is great rain," said Ohlund, looking through the viewfinder as Nancy splashed through the mud with a huge smile spread across her face.

That night at dinner, we learned why the afternoon had been so wet: A tropical depression had gathered just offshore of the Yucatán. Forecasters were predicting it would strengthen into a late season hurricane in the next 24 hours. One of Buddy's employees had printed an Internet forecast suggesting the storm would most likely move north, away from us—although it had a 20 percent chance of moving directly onshore.

Her underwater close-ups completed, Hazel had joined Nancy during a break in the rain. The two had been filming all afternoon, my tourist dive put off for later. Meanwhile, the underwater crew had enjoyed a successful day, unaffected by the rain above. Hazel told us of an unexpected bonus that had come when the team had stumbled upon and filmed a new species of blind underground fish. It appeared distantly related to a type seen only in certain caves of the Bahamas, as described in a single scientific presentation given at an NSS convention Hazel had attended. While she was strictly a microbiologist—and far from expert when it came to larger critters—Hazel knew that her friend

WAKULLA SPRING

BY BARBARA ANNE AM ENDE, PH.D.

Wakulla Spring rises in the Florida Panhandle like liquid crystal flowing amid cypress and live oak trees draped with Spanish moss. Each day a whopping 250 million gallons of freshwater flow up from the spring beneath the earth to form the Wakulla River. Thousands of tourists are drawn each year to swim in the waters and ride glass-bottom boats that allow them to peer into the gaping mouth of the cave that feeds the spring. An abundance of wildlife, including ever present alligators, entertains the visitors on the boat tours.

Relative to the shallow caves in the Yucatán Peninsula, the Wakulla Spring cave system is dauntingly deep—averaging nearly 300 feet below the surface. In the 1950s, scuba diving was a new sport; the equipment was crude yet useful. Two students from nearby Florida State University, Wally Jenkins and Garry Salsman, were curious to learn what lay beyond the entrance of the Wakulla Spring cave. Without pressure gauges for their diving tanks, dressed only in swimming trunks and carrying flashlights wrapped in plastic bags,

the two men dove into the 68° water. They figured that if they had enough air for this dive, they'd know that the next time they went down, they could go just a little farther. Using this simple approach, Garry and Wally penetrated hundreds of feet into the cave—an amazing feat with

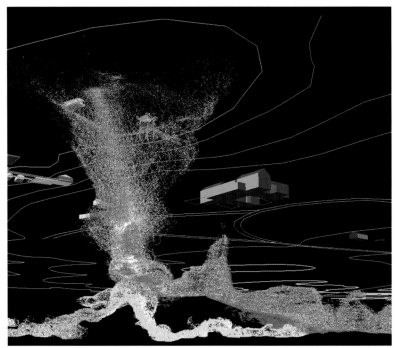

In this perspective view of the Wakulla Cave map, each colored set of points represents a separate data collection run in the cave or surrounding basin. The lines above the cave are topographic contours. Also shown are the State Park Lodge, the bathhouse, and boat dock.

such rudimentary equipment.

No other diving occurred at the privately owned spring until shortly after the land was designated a state park. In 1987 a special permit was issued to the U.S. Deep Caving Team,

Inc. (USDCT) to run an expedition. The goals of the expedition were to map as much of the 10,858-foot cave as they could in approximately six weeks and to test new technical diving techniques. This diving expedition was the first noncommercial and nonmilitary undertaking to use mixed gases in the United States. The problem with breathing only air at deeper depths is the narcotic effect of the nitrogen it contains. Martini's Law suggests that for each 50 feet of depth, a diver responds as though he or she has consumed one martini on an empty stomach. To avoid this nitrogen narcosis, a diver breathes a mixture of helium and oxygen instead of only air.

In addition to pioneering the use of mixed gases for cave diving, the 1987 expedition tested the first prototype of a rebreather, the MK-1, designed by the explorer-inventor Bill Stone. A rebreather is hugely efficient because it recycles a diver's breath. Oxygen is dribbled into the breathing loop at the rate that the human body metabolizes the gas, and the exhaled carbon dioxide is scrubbed out. There are no exhaled

A diver maneuvers the Digital Wall Mapper through a Wakulla Cave passage, taking sonar soundings of surrounding walls. In order to record more exact detail of particularly complex passages, divers sometimes make several passes, resulting in a more tightly clustered "data point cloud."

bubbles and the gas is not wasted. The MK-1 prototype was not ready for actual in-cave use at that time, but it worked well enough to spur Stone to improve the rebreather. Later, in 1995, the MK-4 was successfully used during a USDCT expedition to the Huautla Cave System.

In 1998, the USDCT returned to Florida for the Wakulla 2 expedition. The project was a marvel of modern technology. Not only did the team dive use the now commercially available MK-5 rebreather, but the entire goal of the expedition was to create the first fully three-dimensional cave map. Traditional cave maps are made by surveying a line through the cave and simply sketching the surrounding walls. For Wakulla 2, Stone and electronics engineer Nigel

Jones designed the Digital Wall Mapper (DWM). The front of the unit has 32 sonar transducers spiraled around the nose cone. The sonar measures distances to wall from the mapper. But those measurements mean nothing if the DWM doesn't know where it is when taking the sonar readings. An inertial measurement unit, originally designed for a ballistic missile, is housed in the center of the DWM to track its position and orientation.

All cave maps are subject to error that builds up with each measurement unless it can be corrected. At Wakulla 2, Brian Pease designed special underwater "cave radios" that produced magnetic fields. Divers would place these radio beacons in the cave, then Pease would

locate the signal on the surface 300 feet above and determine a precise GPS (global positioning system) location.

The Wakulla 2 expedition lasted three months and was staffed by 150 dedicated volunteers from eight countries. The result was a spectacular data set of ten million points from the cave walls measured by sonar. The details of the passages were imaged in a way never before possible. Every nook and cranny was precisely scanned. The 3D interactive map allows scientists to compare the features of the cave with surface features in ways never before possible, thereby adding an important tool for the management and understanding of underground water resources.

Jean Krejca, a cave-diving biologist in Texas, was likely to be very interested in the little fish. She was delighted that Wes and Howard had come up with a means of filming the discovery in its native habitat, so that Jean and other experts might identify it.

Two dozen members of the expedition gathered that night in Wes Skiles's cottage at Akumal to view the video Howard had taken along with the day's footage.

"When you're down there in the heat of battle, you don't know what it means," said Wes, his eyes bright. "Take a look at this." He hit the play button, and on a small television in the darkened room we saw a pinpoint of distant light spreading over what appeared to be an enormous, well-decorated cave chamber half full of water in the foreground. The light resolved into divers zooming along behind scooters through the "air-filled" portion of the room—the false waterline below them was the halocline, caught on camera at last. The divers danced in and out of the quicksilver layer, which distorted their features until they wavered like mirages, like holograms on smoke. Here in the stillness of the cave, the undisturbed halocline had been a mirrorlike barrier. Now as each diver passed through a portion of it, the halocline spread into a shiny fog, like a digital image on a computer becoming suddenly pixilated.

"Watch what's coming," said Wes. "You've never seen anything like it." As the camera itself dropped into the ethereal border, Wes let out a deep bass "Wahhooommm." A foot or so below the halocline, the image once more became sharp.

"Zap," he said. "Instant clarity." Wild applause erupted in the room.

The forecasters had been accurate: the storm had moved northeast into the Caribbean, the rain tapering off. While I enjoyed the experience of blasting through jungle atop a Buddymobile, and I had been able to scout a couple of small caves while en route to

A cave diver enjoys a short swim in sunlight, where a collapsed tunnel within the Dos Ojos system has opened a narrow fissure nearly one hundred feet long. Multiple entrances are common in Yucatán caves, where upper-level passages seldom lie more than thirty feet below the surface.

various locations, I spent most of the next few days sitting around watching others work. I began carrying Stephens along to occupy the periods of waiting at cenotes and jungle sets. I sympathized with the explorer when he described the painstaking efforts of his partner to employ the best visual technology of the day in recording the ruins of the Governor's Castle at Uxmal:

> Mr. Catherwood made minute architectural drawings of the whole, and has in his possession the materials for erecting a building exactly like it; and I would remark that, as on our former expedition, he made all his drawings with the camera lucida, for the purpose of obtaining the utmost accuracy of proportion and detail. Besides which, we had with us a Daguerrotype apparatus, the best that could be procured in New-York, with which, immediately on our arrival at Uxmal, Mr. Catherwood began taking views; but the results were not sufficiently perfect to suit his ideas. At times the projecting cornices and ornaments threw parts of the subject in shade, while others were in broad sunshine; so that, while parts were brought out well, other parts required pencil drawings to supply their defects. They gave a general idea of the character of the buildings, but would not do to put into the hands of the engraver without copying the views on paper, and introducing the defective parts, which would require more labour than that of making at once complete original drawings.

In order to get out and see more of the country, I began volunteering to run an assortment of errands—ferrying crew members from one location to another, picking up new arrivals from the airport in Cancún, driving to town for everything from dry ice to turista remedies. What would appear as a single, uninterrupted cave dive in the finished film would actually be shot over several days in a variety of area cenotes, and I enjoyed scouting these often remote locations with the topside crew in advance of each day's shoot. One errand brought me to the offices of the Centro Geologico Akumal, a small scientific agency overseen by Charles Shaw, a geologist who has spent the past two decades studying the peninsula's connection to the death of the

An unexpected bonus to the expedition came when Hazel stumbled upon a new species of blind underground fish. It appeared distantly related to a type seen before only in certain caves of the Bahamas.

dinosaurs at the end of the Cretaceous age.

After the storm had moved off to the north, Jack had bolted an IMAX camera to the wing of a chartered plane, recording aerial views of Tulum and the surrounding jungle. He also photographed a second plane as Nancy scouted cenotes by air. But a particular rumored cenote—a mysterious blue circle over a hundred feet in diameter, surrounded by thick jungle on all sides—had eluded them. They had been given rough directions penciled onto a highway map, but from the air no one had been able to spot the large pool.

I agreed to try to locate a topographic map of the area, which should have sufficient detail to lead directly to the cenote. From the small natural history museum at Akumal I was directed to Shaw's cluttered office. He was dressed in faded khakis, his unkempt gray hair framing a thoughtful lined face, and blue eyes that seemed to shine with intelligence. When I described what I was after, he pulled a yellowing map from one of a dozen boxes of charts and unfurled it on his desk. The Akumal resort, which had opened in the 1950s, was a cluster of seaside dots on the map; none of the larger resorts that now hugged the coast had existed when it was printed.

"I know every ridge and cenote out there," he said. "I suspect this is the one you had in mind." He pointed to a neat circle about two miles to the west of an indicated bend in the power lines that followed the coast south from Cancún—ten miles or more from the spot where Jack had searched. It was unquestionably the large cenote called for in the script. Shaw invited me to borrow the map, provided I promised to get it back to him. "That series is long out of print," he said.

I asked him about his work, and he explained that his current interest was the dangers posed to the aquifer by increased development. Wastewater from the coast was seeping inland at alarming rate, he said. He was working with researchers from Mexico and several American universities to document the extent of the pollution and, if possible, to create a plan to alleviate it. Almost as an afterthought, he explained that he was very familiar with groundwater flow across the Yucatán because of his work with Chicxulub.

"Sheek-shoe-lube?" I asked.

"Chicxulub," he repeated, and wrote the word on a sheet of scrap paper. "It's a little village on the north coast, near the impact center of the meteorite that killed the dinosaurs."

CAVING FOR TROGLOBITES

BY JEAN KREJCA

The three of us were huddled in a warm, humid room about the size of the interior of an average two-door car. We had crossed an international border, come many miles over rough terrain, and had finally rappelled down 215 feet of vertical cave to reach this isolated, muddy room and a pool that would allow me to explore into underwater passage beyond. I slid into the water from the slick clay slope, causing silt clouds to barrel downslope into the already milky water. My field of vision instantly changed. Now I was enveloped in bright diffused washout, a picture as useless as a snow-filled television screen. Finally the passage opened up and visibility improved, and *Yes!*—a blind catfish, what sweet success.

Following it, now dropping down, I could easily identify the catfish as an individual we had marked some six months prior. I ascended into a mud-covered cave, which meant that at times murky floodwaters entirely filled this underground chamber. The air did not move; not a trace of surface input such as a twig or a rotten leaf, or even a cricket or beetle, ubiquitous in the rest of the cave. This is the remote, sunless habitat where troglobites live. Rarely do humans reach these underground caves—the conduits in which both obligate cave organisms, or troglobites, travel freely and the aquifer water we rely on flows.

Like many cave scientists, I began with a Huck Finn–type of love for exploration. But the thrill of cave exploration often ends in spots that are just too tight or otherwise difficult to pass through even with specialized equipment. My goal is to prove that exploration through these inaccessible cracks can continue by tracing troglobites, tiny cave invertebrates that travel there, through genetic histories and thereby mapping out their offspring and the different caves and aquifer regions in which they move.

It is essential to map groundwater regions, a costly and difficult endeavor. Without such charts, it is nearly impossible to project the impacts of contamination and effectively manage them. It is imperative that we trace underground water passages and determine their paths and contaminant spread, how and where the pollutants move through the aquifer. Hydrologists traditionally use techniques such as dye tracing, potentiometric surface mapping (water table-level maps based on data from wells), and mapping of geologic units and caves to estimate how, or if, aquifer regions are connected. These maps give important data but are rarely detailed enough for a given region. A mapping method that has not been attempted, which I am currently testing, tracks troglobites, specifically the aquatic crustaceans that live, reproduce, and travel in the watery underground regions of our planet.

There are numerous ways to track organisms. The blind catfish mentioned above was part of a mark-and-recapture study. Physically marked with a tattoo, we will be able to determine its movement. Currently, however, genetic markers appear to provide the longer range and more reliable trace because troglobites may not move very far along the aquifer superhighway: they are extremely patriotic. By using genetics—analyzing DNA and diagramming trees of evolutionary history—generations of these aquatic critters will be traced over longer periods of time and perhaps over greater distances. Crustacean cousins will carry the genes of an ancestor that was sampled in another region of the aquifer. Ultimately, the genes of troglobites collected may indicate the conduits in which they traveled and thus the connections between various hydro-geological regions.

My work specifically focuses on tracking and collecting troglobites from a number of cave and well sites across the Edwards Aquifer of Texas and Mexico. If the evolutionary trees for the troglobites living in these underground habits are compared to evolutionary trees resulting from traditional hydrological mapping methods, and the trees are congruent, there will be corroborating evidence that underground tracking of aquatic crustaceans corresponds to the hydrological connections. The next step will be to develop this into a standard mapping technique used in karst, predominantly limestone, aquifer management. Then when I once again reach an underground passage too narrow to explore and continue sample collection, I will be not be disappointed: troglobites travel where humans cannot.

I was vaguely aware that there had been growing scientific acceptance of the once controversial theory that a massive meteorite had ended the reign of the dinosaurs; I had read that the likely impact site lay in the Gulf of Mexico. But, as Shaw related, the massive explosion had made it possible for the Yucatán to become unique among karst regions as the land of cenotes. He pulled a sheet of printer paper off a nearby shelf. It depicted the Yucatán as seen from space, in the false colors of a satellite radar image. Cenotes that looked randomly distributed from the jungle, and even from the air, fit a clear pattern when seen from orbit.

The solid block of limestone from which the Maya world had grown began as a seafloor deposits upon the shocked and fractured surface of a vast crater created 65 million years ago. Concentric shock rings surrounded the impact site like the colored bands of an archery target. When the limestone that accumulated atop these features was lifted some 50 million years later, a pattern was already in place for the growth of later caves. In the image from space, the blue dots representing cenotes formed thick bands around two major rings of the buried crater. Those cenotes not directly on the rings were more sparsely distributed and followed no pattern that I could see, but Shaw said that even they could not have formed so uniformly throughout the region had not the ancient disaster dictated the deep, even placement of the aquifer.

"This aquifer is a great natural resource, unique in the world," he concluded. "And it is increasingly under threat."

I thanked him once again for the map, and returned to my rental car to begin the now familiar journey down the rough limestone road to Dos Ojos. As I drove, I tried to envision the planetary calamity the rock concealed. I was finally becoming eager to see this aquifer from the inside, Hazel had said she was free from science and filmmaking in the morning, a certified instructor had given his blessing and agreed to guide, and we had planned to meet at the dock at nine. I was going diving.

Buddy Quattlebaum kayaks over the calm surface of a stadium-size cenote cavers reach by rope ladder, much as ancient Mayas did in times of drought. Cave divers often lower boats into larger, deeper sinkholes to serve as portable dive platforms.

CHAPTER 8

INCIDENTS OF TRAVEL

Imagine flying weightless through a blue palace, surrounded by glittering jeweled curtains and columns. Streaks of distant daylight filter through the chamber from some unseen place behind you. There are no sounds save the hiss of in-drawn breath and the gurgle of exhaled bubbles. The warm water enfolds you like a pleasant bath. The sides of your face mask, blocking peripheral vision, frame the magnificent scene before you. ■ When you turn your head, your eyes register the lensing effect of the water, the dreamlike skew of photons passing through a medium other than air. Your handheld flashlight illuminates calcite formations, some a brilliant white, others a mottled yellow or orange. On the uneven floor ten feet below lie the scattered bones of ice age animals, dead since the time 10,000 years ago or more that this room held air. ■ I was beneath Dos Ojos, and at last I began to understand the lure of cave diving. The well-decorated passage before me resembled many caves, in size and shape, that I had explored. Now, instead of scrambling over muddy slopes and dusty rocks, I flew effortlessly through the center of the tunnel connecting the two air-filled chambers that surround the sinkhole. Although I remained conscious of the air gauge hanging closely at my side, of the ceiling overhead, and of the nearby location of daylight and air, I could sense the seduction that had drawn so many untrained open-water divers to their deaths in caves. The submerged

Above: Hazel prepares to begin a dive in Dos Ojos using a small electric scooter, useful in conserving air by rapidly transporting her. Right: Scooters pull two divers into a deep, massive chamber known simply as The Pit.

cavern was like a sensory-deprivation tank, removing touch, sound, and gravity, forcing my mind to focus on the beautiful images unrolling before me. It was wholly unlike the intensely physical activity I thought of as caving; it was like watching a movie of caving. I could sense how easy it would be to become distracted by the show, how surprising it would be to experience a sudden collapse or silt-out that demanded active participation.

Fortunately, such an event was highly unlikely on this excursion. My fingers loosely circled a taut nylon dive line as I followed Hazel and a commercial dive guide named Greg Brown, yet another expat American drawn south to live among the cenotes, along the main tourist route around the perimeter of Dos Ojos. According to the depth gauge I carried, I was 30 feet down. Only water and rock stretched overhead, but air and daylight

lay where I could swim to them on a single breath, should I feel the need to do so. The passage was so large that I doubted an Olympic synchronized swim team could stir up enough silt to obscure it. The buoyancy compensator I wore as a vest raised and lowered me to whatever vertical level I desired.

Hazel and I had done a buddy-breathing exercise at the dock and practiced the various cavern-diving hand signals. Greg was satisfied that I could handle a route where the only real danger that could arise would have

to come from me, in the form of sudden and irrational panic. Far from panic, I felt an almost narcotic euphoria as I followed my guides through the otherworldly realm. I understood why Hazel had insisted I experience a small part of the cave as she and her more seasoned colleagues saw it. Neither my fin movements nor my finessing of the BC had the ease and grace displayed by Hazel or Greg, yet compared to the grunt work of caving as I had known it, this was like gliding by thought alone. Here was the path the Maya gods had followed from the underworld to the surface. I felt for a moment as though I flew beside them.

We paused above a small depression, where Greg showed me the skull and bones of a small jungle cat, hidden away from where any souvenir-seeking tourists were likely to spot it, then we rounded a large boulder and I saw that we had circled back to the main dock, where once more Howard was reloading the underwater camera for a shot in some distant passage. I surfaced, grinning.

Hazel turned toward me as she climbed the ladder to the dock. "Ya see what I mean?"

"Absolutely! That was stunning," I exclaimed. "Absolutely wonderful."

I retained my resolve never to take up serious cave diving, but a guided cenote tour was something else again. It offered a controlled—and above all safe—taste of the sport's allure, of the beauty of the Yucatán's aquifer. My entire trip had taken less than 30 minutes. From it I gained a new respect for those divers who kept their wits and training about them for hours while exploring virgin passage thousands of feet from any entrance. And I found myself feeling fresh sympathy for the unfortunate dead who had failed to do so.

That afternoon I joined Hazel, Nancy, and the topside crew at an air-filled cave located down a couple of miles of horrendous road from Dos Ojos. I had foolishly tried to reach the spot with my compact rental car; after

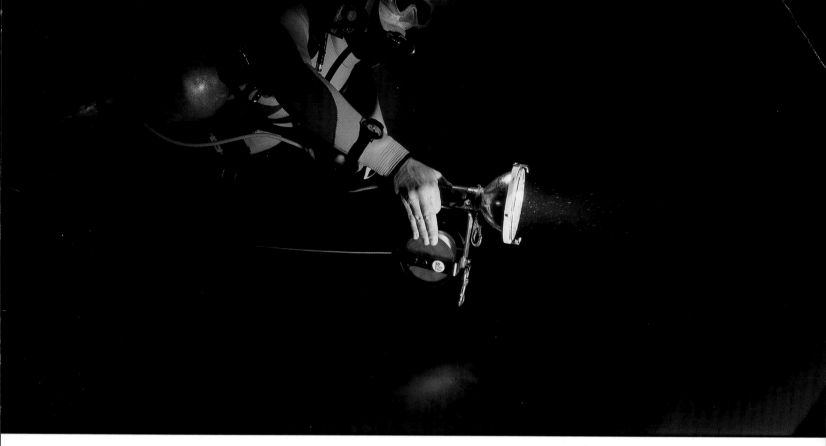

bottoming out for the third time in a hole that could almost qualify as a cave, I squeezed into a narrow spot off the road, out of the way of the Buddymobiles, and took off on foot. The main passage of the cave was nearly a hundred feet wide and 30 feet tall and sported a forest of thick calcite columns near the wide entrance. Although it contained several sandbars and dry side tunnels, most of the cave lay under two to three feet of water, with a much deeper pool located near a rear entrance perhaps 500 feet down the main passage.

When Steve Judson had first scouted this location with Buddy and Manuel, they had found fresh jaguar prints in the soft mud, evidence that the great cat of Maya lore has not quite abandoned its native terrain even though it has been hunted to near extinction. For the purposes of the film, the place had thus been dubbed Jaguar Cave. With the loud activity of a film crew crowding the entrance area, complete with gasoline generators, stadium-size lamps, fog machines, and tangle of cables thicker even than the jungle lianas hanging from every tree, it seemed unlikely that a jaguar would wander within 20 miles of the place. But my awareness of the animal's recent presence, together with the shards of Maya pottery that I soon spotted in a dry alcove just inside the cave, once more plunged me into the Yucatán of Stephens and Catherwood.

I recalled one of the caves Stephens had vividly described. In the winter of 1842, he and Catherwood arrived in the Maya village

Continued on page 126

The dive line, kept spooled on a hand-held reel, is more than a safety device: When marked with distances, it also allows divers to measure passage length. Each tie-off point is recorded on a waterproof slate, so that the diver can calculate the total passage traversed.

121

A team member
steers a scooter along
the stalactite-draped
ceiling of a Dos Ojos
chamber. Unlike open-
water divers, who carry
their air on their backs,
most cave divers wear air
tanks mounted on their
sides in specially made
harnesses. This creates
a lower profile, making
it easier to negotiate
tight squeezes.

CAVE OF THE SNOTTITES

By Louise D. Hose, Ph.D.

The ultimate thrill for any explorer is venturing into a place where no one has previously gone. As of January 2000, Cueva de Villa Luz in the southern Mexican state of Tabasco retained one known unexplored area: a stream passage called Other Buzzing Passage, extending beyond the Lake of the Yellow Roses.

Jim Pisarowicz checked a low passage leading off to the right and I checked one to the left. Crawling on his back in order to keep his gas mask filters out of the stream, Jim returned after only a few minutes with his report. "I had to stop when my filters became clogged with the gypsum paste from the low ceiling."

My motivation to push it any farther quickly waned: the tight passage ended with a chimney-size vertical void overhead. I squirmed into position and started climbing. The walls were mush. Cavers commonly wedge their way up mud-lined climbs, but these walls were different. The coating was not mud but the same gypsum paste Jim had encountered: a white, toothpaste-like substance rich in acid-secreting bacteria, having the corrosive strength of car-battery acid. It was a notably unpleasant substance to wallow in.

We already knew Lake of the Yellow Roses to be a unique area in one of the world's most bizarre caves. Undulating layers of sulfur coated the walls above the small pool and high concentrations of hydrogen sulfide (the rotten-egg gas) in the room's atmosphere

forced us to change gas mask filters every half hour.

When members of our team made the first penetration ever into this area of the cave, far from the entrance, they reported an additional smell, similar to ammonia, which we identified as another microbial emission—formaldehyde. Each trip into the room

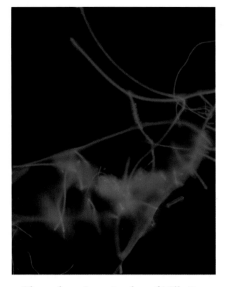

The rock-eating microbes of Villa Luz provide food and habitat for cave-adapted insects and spiders, which in turn feed larger organisms.

brought a new respect for the demands of the environment. We had learned not only to use gas masks but that we also needed to carry our National Geographic–funded electronic monitors to measure gas concentrations. When Jim and I had previously tried to explore the area beyond Yellow Roses, the gas monitor screamed out its alarm. Hydrogen sulfide concentrations were elevated, along with carbon monoxide and carbon dioxide. But

more distressing, the oxygen level had dropped from its usual 20.9 to 9.5 percent. Since that trip, we had carried miniature scuba tanks on our belts. The gas mask filters can remove the hydrogen sulfide, formaldehyde, and some other gases before reaching us—but they will provide no protection against lethal levels of carbon dioxide or insufficient oxygen.

The idea of chemoautotrophic ecosystems in caves, where microbes make a living off of the energy released by chemical reactions with sulfur and other nonphotosynthetic processes, was in its infancy when Jim Pisarowicz first wandered into Cueva de Villa Luz in 1984. A psychologist by training and a devoted caver, Jim immediately recognized Villa Luz as a scientific treasure trove—"I knew something was going on in this cave, and going on in a big way!" Pisarowicz later enthused. Three features of the cave drew his attention: 1) the high hydrogen sulfide concentration in the atmosphere; 2) a remarkable abundance of small, pale pink fish in the stream; and 3) rubbery stalactite-like microbial deposits that dripped concentrated sulfuric acid, which he dubbed "snottites."

Dozens of springs rise into the air-filled cave at the heart of the extreme environment of Cueva de Villa Luz. The water in the springs rises from deep in the Earth, probably from a nearby oil field. The water is highly charged with various dissolved gases including hydrogen sulfide, carbon monoxide, and probably methane. All of these

Caver Jim Pisaowicz gave these acid-secreting microbial strands the ignominious—but apt—name of "snottites."
The acid at the tips of some snottites has been measured at incredible 0.0 pH.

gases, unstable in the presence of free oxygen, will rapidly convert to another substance when exposed to O_2. Hydrogen sulfide (H_2S) changes to sulfur dioxide (SO_2) when combined with oxygen and to sulfuric acid (H_2SO_4) when exposed to oxygen and water. Thus, when the H_2S is carried into the oxygen-rich, humid, air-filled cave, the gas combines with O_2 and H_2O to form sulfuric acid on the cave walls. Because this chemical reaction releases energy, microbes participate in the process and use the energy to build cell structures—just as other plants use sunlight in their process of photosynthesis. In a similar manner, microbes help convert methane to formaldehyde and carbon monoxide to carbon dioxide as the gases transit from the oxygen-starved deep plumbing system below into the oxygen-rich, air-filled cave.

The microbes form the base of a complex and robust food web in the cave. Slimes coat the walls throughout the cave and snottites are suspended above the stream. Some springs belch out "stringers," one- to two-inch-long strings, and "phlegm balls," one- to two-inch-diameter disks reminiscent of oysters—all of which are entirely consisting of microbes. Bacteria play the same role in this subterranean ecosystem as plants do on the surface: abundant midges and fish eat the microbes; in turn spiders, bugs, and fish consume the midges: there is so much energy that all can feast and abundant food supplies continue. Zoques, the indigenous people, regard the cave as sacred, confident that the cave fish will provide sufficient food even in times of drought or crop failure. The springs in the cave always flow.

Deeming the exploration to the end of the Other Buzzing Passage finished, Jim and I started a detailed survey and mapping of the passage while returning to Yellow Roses. I felt relief and comfort that I was leaving this most unfriendly of cave passages. In such a frame of mind, I reluctantly checked the last low crawlway in the area.

The deeply etched, bare limestone floor lacked the usual slimes or sediments found everywhere else in the cave. The gas monitor screamed as I crawled on my belly; I switched to the scuba tank, uncertain that the partially used gas-mask filters provided enough protection. Another spring erupted in the floor at the end of the crawl. I named it Pozo Obscuro, the hidden well. Laboring to return with testing equipment, I quickly measured the new spring's water. It was the most unusual yet. Like most interesting cave discoveries, Pozo Obscuro revealed itself on the last day of our NGS-sponsored expedition. We dragged ourselves out of the cave, removing any trace of our presence, and plotted our return next summer to continue the scientific exploration of one of the world's most extreme cave environments.

of Bolonchén de Rejón, which according to Stephens translates as "nine wells."

"From time immemorial, nine wells formed at this place, the center of a sizeable population, and these nine wells are now in a plaza of the village," he wrote. "Their origin is as obscure and unknown as that of the ruined cities which strew the land, and as little thought of." Stephens explained that even with diligent care and upkeep by the village authorities, the wells provided water for no more than seven or eight months a year. As the rainy season drew to a close in the fall, the time approached "when these wells would fail, and the inhabitants be driven to an extraordinary cueva at half a league from the village." On an appointed day the village would hold a feast at the cave's entrance. A party of men would work to repair the many ladders that led to the cave's remote pools, from which jugs of water would be hauled daily until the rains returned.

Stephens and Catherwood set out to visit the cave, now called the Grutas de Xtacumbilxunaan, working their way to the base of a steep sinkhole and through the rocky entrance by torchlight. They climbed down a 20-foot ladder to a short passage that ended at a balcony in the side of a vertical shaft 210 feet deep. A small overhead connection to the surface provided a skylight to illuminate the scene.

"From the brink on which we stood, an enormous ladder, of the rudest possible construction, led to the bottom of the hole." Stephens wrote that the ladder was between 70 and 80 feet long and 12 feet wide, made of rough tree trunks and saplings lashed together with vines. Their local guides began descending the precarious-looking stair, "but the foremost had scarcely got his head below the surface before one of the rounds slipped, and he only saved himself by clinging to another." Nervously, Stephens and Catherwood worked their way downward. By keeping each hand and foot on a different log rung, with "an occasional crash and slide," the party reached the bottom.

At the base of the pit Stephens and Catherwood encountered an entrance to another downward sloping passage, where once again the party found rickety ladders, although none so long or steep as the first. Eventually, by using a rope in places, they were able to work their way down to an intersection of branching passages. Two hundred feet down one of these tunnels, they descended a ladder to "a low and stifling passage; and crawling along this on our hands and feet, at a distance of about 300 feet we came to a rocky basin full of water. Before reaching it one of our torches had gone out, and other was then expiring." By Stephen's best calculation, they were 450 feet below the surface and 1,400 feet horizontally from the entrance.

The two explorers rewarded themselves with a bath in the red-tinged pool, the first they had taken after days in the dry countryside. Eventually they visited several other chambers containing pools, each of which had a different color and clarity. For four or five months of the year, these deep pools provided the only water source for the village, just as now cities and 5,000-room resorts on the Yucatán are watered by larger, stronger rivers that course below the peninsula. Forming complex systems hundreds of miles long, these hidden streams are now being explored for the first time.

I once saw Catherwood's painting of Maya carrying jugs up this precarious structure in a geologist's slide show on the management of karst. George Veni, an expert hydrologist, had paused on the slide to make the point that all people living on the classic karst landscape, which is dominated by limestone, drew their drinking water from caves, whether they were aware of it or not. His next slide showed a gently rounded hillside marked by the precise square pits of an archaeological excavation. Veni explained that the dig site was a trash dump and common sewer of the ancient village of Bolonchén de Rejón, situated in the recharge area directly above the cave. Rainwater leaching through generations of

Helped by resident Maya guides, Nancy Aulenbach bushwhacks to a cave entrance to ferry supplies to emerging divers. "Dry" cavers like Nancy often volunteer to work on surface support teams for cave divers.

Nancy takes advantage of a high-wattage movie lamp to examine formations on the ceiling of Mil Columnas, the cave of "a thousand columns." The cave's shallow stream was littered with pottery sherds—testimony to the cave's long use by the Maya, who constructed an ancient altar near its deepest pool.

the village's refuse would have flowed directly into the pools from which the town drew its sustenance.

"It's a safe bet that the inhabitants of this village suffered chronic gastrointestinal disease," the geologist had said. "Anyone living on karst should keep the relationship between their sewage and their drinking water in mind when they set out to build."

I had recalled Veni's warning when I read in Stephens's account that the day after his bath, he and other members of the caving

party came down with fevers that, together with jungle fleas, tormented them throughout the following night. Unaware of any virulent organisms that may have lurked in the hidden pools of Xtacumbilxunaan, Catherwood and Stephens had been caught up in the excitement of exploring the deeper passages. "The cave was damp, and the rocks and the ladder were slippery," Stephens wrote of his precarious journey to the lowermost pool. "At this place the rest of our attendants left us, the village ministrio being the last deserter. It was evident that the labor of exploring this cave was to be greatly increased by the state of the ladders, and there might be some danger attending it, but, even after all that we had seen of caves, there was something so wild and grand in this that we could not bring ourselves to give up the attempt."

It was a feeling I knew well.

Not far from the main entrance of Jaguar Cave was an overhead skylight, 25 feet or so above a dry shelf beside the main passage. While the film crew began the complicated procedure of dropping electrical cable and rigging lighting through this skylight, I took the opportunity to explore the cave. The three side passages I located were short, generally decreasing from easy walking height to stoop-walks to hands-and-knees crawls before petering out altogether.

Unlike the main passage, they were not well decorated and at first glance were unremarkable. However, having time to kill, in each one I carefully looked for flat shelves or alcoves, which I found several times near the ceiling. In nearly every case, close inspection of these shelves revealed broken bits of pottery, some incised with decorations. I guessed that these were the remains of tallow lamps, or perhaps broken water jars. It would be up to some future archaeologist to determine their purpose; I left them undisturbed.

Near the rear entrance, a stone altar about ten feet square had been constructed at water level near a deep blue pool—the only spot in the cave that appeared deep enough to possibly offer connection to the aquifer. While the unadorned altar may have been first constructed when the pottery sherds were fresh, its use was by no means archaic: I had learned that a local shaman still performed rites in this cave. Sometimes versions of certain ceremonies would be conducted for the benefit of the hardier ecotourists that Buddy occasionally brought here. Other, more secret rituals took place in private, among those few Maya who sought to keep their culture alive.

Extending out over the deeper water was a rounded natural shelf about six feet long, three feet wide, and eight inches thick, just below water level. I recognized this as "shelf-stone," a type of calcite formation common in the pools of Lechuguilla Cave in New Mexico. The shelf would have begun as a wafer thin calcium raft, floating upon the surface of a pool remaining undisturbed for years at a time.

When one side of the raft cemented itself to the rock at the sides of the pool, it provided enough stability for the layer of mineral deposit to thicken without sinking to the bottom. Over time—centuries, at least—the raft had accreted sufficient mass that it was now a solid shelf, capable of supporting the weight of several people. By walking closer to the back entrance of the cave, where the water was only inches deep, I was able to spot a series of puddles where the process seemed to be just beginning.

The basic chemistry of such accumulation is well understood: when a single microscopic crystal of calcium carbonate precipitates from water saturated with the mineral, it forms a nucleus that tends to grow for as long as the site comes in contact with the saturated water. A tiny soda straw formation thus builds great stalactites and columns. What is not well understood, and has become a controversial topic among cave scientists in recent years, is the role that life might play in the process.

Geologists who study travertine dams—the aboveground cousins of cave formations—have found that the initial precipitation often occurs around masses of bacteria, which are preserved as fossils in most travertine formations. Some have theorized that without the bacteria acting as a catalyst to precipitate calcite from the surrounding water, the mineral formations would not form at all. Inspired primarily by the microbial wealth of Lechuguilla, and the many ways microbes seemed intertwined with that cave's geology, a few cave microbiologists have begun gathering evidence that

certain cave formations are not only the result of chemical reactions, but are signs of biology in action.

"Behind every great cave formation is a great microbe," claims one microbiologist I know. Skeptics argue that while living and fossilized microbes have been found within many types of cave formation, the microbiologists have not yet proved that bugs are what cause the formations to exist in the first place. "It may be that they just like certain real estate," counters one such doubter.

Here in the Jaguar Cave, I did not see much room for doubt. Just inside the rear entrance, a thick organic scum floated atop wide puddles of tepid, undisturbed water. If I were to step into one of these, scum would swirl and vanish. Yet as I lifted a bit of it with

a twig, I could see that small white dots strung with filaments—suggestive of certain bacterial colonies and yeasts—sparkled with tiny grains of what appeared to be calcite. The crystals and the dots were definitely related to each other; nowhere in the pond could I find one without the other.

In places, enough of these grains had joined together to form translucent circles the size of coins. These were far more mineral than organic in appearance. I suspected that, left alone for a century or two, they would prove to be the nuclei of new calcite dams and shelfstones. Of course, actually proving such a thing might be the topic of a doctoral dissertation project. Without such proof, my speculation would have to remain just that. Still, I resolved to ask Hazel to devote one or two of her sample collection bottles to the scum.

I stepped out of the back door, climbing out a sinkhole lined with thorns and muck into a sun-dappled forest. Surprisingly, I could not hear the generators nor any trace of the film crew, even though I knew they could be no more than a few hundred yards away. Looking behind me, I saw that the cave had penetrated a low overgrown ridge; both jungle and rock shielded me from the main entrance area. I could be no farther from the rest of my party than the length a football field. Had I chosen to run back through the cave, I could have reached the others in less than five minutes. Yet here on the surface I felt utterly alone.

The ridge reminded me of the limestone bluffs common to certain swamps of north Florida, where I first began caving. There, the best place to find new cave entrances was by walking along the base of a bluff near a floodplain. I could see that just beyond the sinkhole I had climbed lay the base of the ridge. The land beyond it appeared appropriately low and wet. I began working my way along the ridge, slowly and carefully moving dead branches and hanging vines from my path. I suspected that here, as in Florida, the cool stone might offer refuge from the heat to poisonous snakes. I had no desire to meet up with a fer de lance—a Central American viper with a bite far deadlier than the rattlesnake's. Twice I saw the shiny red bark of a tree that Manuel had warned would cause painful blisters if it touched the skin; I gave it a wide berth.

I worked my way through the vicious vegetation overland toward the main entrance, a cloud of insects hovering about my sweating face. Each armored branch that I pulled aside revealed another ten just like it. When, after 15 minute's slow progress, I could still not hear the generators, I began to fear that I had somehow misjudged my location, was actually fighting my way deeper into the bush. Although I had mentioned to Nancy that I was going to poke around the cave a bit, no

one knew that I had left it. I was on the verge of feeling extremely stupid when I stumbled out of particularly thick clump of prickles and heard the hum of the generators, more or less where I had thought they should be.

In ten minutes I was back at the main entrance, sweaty and relieved. I paused as Tom and the others filmed the principal cavers chopping a path to the cave from the undisturbed jungle beyond the entrance. After the sequence was completed, I went back into the main passage to wash off in the cave stream, savoring the feel of the cool water after my hot and uncomfortable hike. While the next shot was arranged, I showed Hazel my calcite pond scum; she agreed that it might be interesting to sample. Nancy showed me a layer in the columns and rock walls near the ceiling that was a clear high-water mark of long-standing. She suggested it may have represented a past sea level: coastline of the Yucatán has wandered as

much as a hundred miles as changes in the Earth's climate over the past few million years have built up and reduced the polar ice caps.

One of the scenes to be shot in a topside cave was of Hazel placing sample vials from the halocline into a canister of liquid nitrogen, in which the specimens would be

Hazel collects a biological-looking scum from the surface of water near a horizontal cave entrance. In order to preserve DNA for later genetic analysis, she often carries a metal canister of liquid nitrogen into the field, which can keep samples below freezing temperatures for up to a week.

preserved until they reached Hazel's research lab. Unfortunately it turned out that liquid nitrogen was impossible to come by in Cancún, or anywhere else on the peninsula. One of the associate producers had managed to find some dry ice, which would steam appropriately for the camera but would melt long before the sample reached the U.S. Hazel informed me that she would return home to the Pace Lab with great memories and terrific footage, sans specimens.

"Not to worry, though," she said with a wink. "This just means I have to come back down here in a couple of months with a good supply of nitrogen, do all the diving again and get some fresh samples. I couldn't have planned it any better."

For now, though, the halocline and the caves the team had explored would have to keep their microbial secrets. The two cavers and I returned to just inside the entrance area, where the crew was still setting up the next shot inside. I sat down on a sandbank near some of the Maya porters and absent-mindedly raked my hand through the gravel at my side. When I looked into my hand, there rested a triangle of pottery, smooth on one side, lined on the other with indentations clearly made by the potter's fingernails. Pieces to the puzzle of past life were all around me.

Science, I knew, was the long process of putting the puzzle together in full knowledge that a complete picture might not emerge for generations. Although genetic analysis of new organisms and environments had advanced biology tremendously over the previous decade, the microbial analyses of the Earth's extreme environments was a science very much in its infancy. Only a little more was known about subsurface bacteria in the Yucatán than was known about subsurface life that might exist beneath Mars or Europa. The thrill of such of a young field of investigation to researchers like Hazel was that so much new life was out there, waiting to be found. The frustration was that years might pass before they could know the true significance of any of their finds.

Hazel, front, and Nancy, use a kayak to travel through Mil Columnas the easy way.

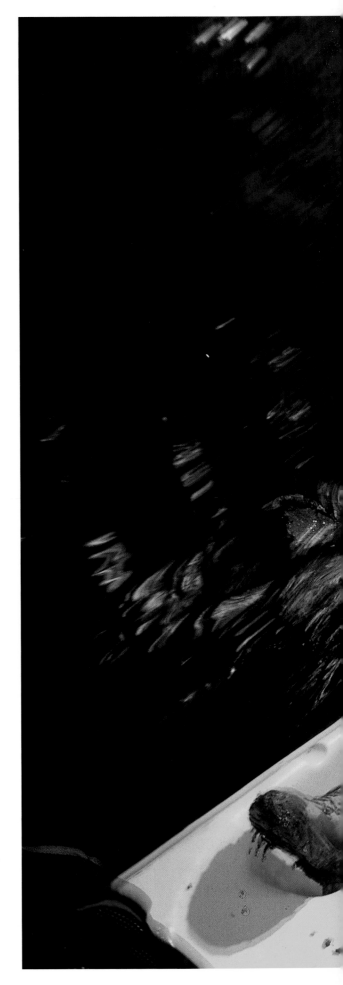

EARTH

PIT BOUNCING AND CRAWLWAY WRIGGLING

A

T LAST, THE STORM
CEASES, AND WE GO ON. WE HAVE
CUT THROUGH THE SANDSTONES
AND LIMESTONES MET IN THE
UPPER PART OF THE CANYON,
AND THROUGH ONE GREAT BED
OF MARBLE A THOUSAND FEET
IN THICKNESS. IN THIS,
GREAT NUMBERS OF CAVES
ARE HOLLOWED OUT,
AND CARVINGS SEEN, WHICH
SUGGEST ARCHITECTURAL FORMS,
THOUGH ON A SCALE SO GRAND
THAT ARCHITECTURAL TERMS
BELITTLE THEM.

John Wesley Powell
in Diary of the First Trip
Through the Grand Canyon, 1869

*Previous Pages: The flame of caver Chris Stine's carbide
lamp becomes a yellow smear as he climbs up a typical
formation-bejeweled passage of Lechuguilla Cave in this
timed exposure. Right: The narrow canyon of the
Little Colorado River snakes through the Navajo
Reservation to join the Grand Canyon of the Colorado.*

Above: Light shines through translucent soda straws and a stalactite, all made of white calcium carbonate tinged red by iron and other minerals. Right: Using the single-rope-technique developed in TAG, a caver rappels into a lower-level passage of Arkansas' Blanchard Springs Cavern.

TAG caver Jim Hewett uses mechanical ascenders to climb into a passage he discovered and named "The Most Horrible Thing Ever." With water and mud common companions to vertical caving, cavers depend on metal-toothed gear that can cling to ropes so slick that even Tarzan would be hard pressed to climb them bare-handed.

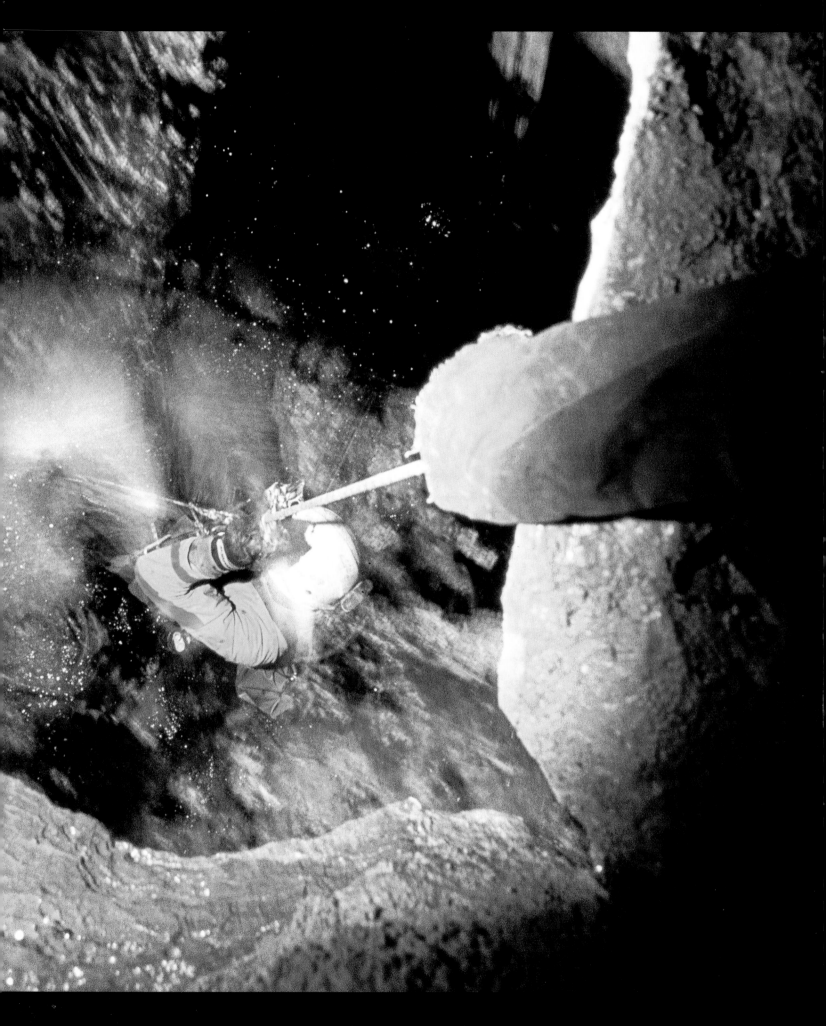

CHAPTER 9

NO CAVES TO SPEAK OF

On a dry afternoon in the summer of 2000, with the temperature hovering at 112° F, a group of cavers and filmmakers stepped from a helicopter as its blades fought for purchase in the overheated air. They eased their way down a crumbling spine of red rock toward the edge of a cliff 2,000 feet above the Little Colorado River, just upstream of its junction with the Colorado on the western edge of the Navajo Reservation in Arizona. The confluence marks the end of Marble Canyon and the beginning of the Grand Canyon of the Colorado. ■ Halfway down the sheer face of Redwall limestone lay the entrance to a virgin cave, visible for centuries from the canyon floor yet never entered because of its inaccessibility. Nancy Aulenbach would be the first one down, rappelling 300 feet and then swinging inward toward the large recessed opening 350 feet above the next ledge, the milky blue rapids of the Little Colorado streaming beneath her feet far below. The day before, she and Hazel Barton had shot through these rapids on kayaks, following the lead of local guides as they examined dozens of mineral springs bubbling into the river—the outflow from hidden caves. The whitish tinge of the river came from dissolved calcium carbonate in these springs, released from the limestone walls of growing caves. Upon exposure to air, the calcium carbonate quickly became solid once more, accreting in colorful travertine dams that hid trillions of microbial fossils.

Above: Despite the loss of an arm in the Civil War, Major John Wesley Powell made the first descent of the Grand Canyon in 1869. Right: In May 2000, Hazel descends to a cave never before entered, 2000 feet above the Little Colorado River.

Despite the heat, despite the inspiring yet terrifying view from the edge, Nancy calmly geared up, taking her time, familiar at last with the slow process by which caving was recorded on film She didn't begrudge exploring with a film crew—the solid metal framework they had built out over the lip would provide an ideal rig point. The platform housed winches that could raise and lower the camera and crew members to exact points along the canyon wall. It was the brainchild of Earl Wiggins, the rigger, whose expertise has been applied to numerous other famous Hollywood moments, including the rigging used by Tom Cruise for the dramatic climbing sequence that opened *Mission Impossible 2.* But here on a cliff in the Grand Canyon, the level of vertical exposure that cavers and film crew would face was even greater than the vertical exposure from which Wiggins had suspended Cruise.

It wasn't every day that a TAG caver got the chance to enter a large canyon entrance never explored by the locals, let alone with professional rigging that could have lifted a horse to the cave. Arizona cavers were notoriously secretive with cave locations, seldom revealing information to cavers from TAG, or anywhere else. If movie making was the price of admission into a virgin Arizona cave on such a grand scale, Nancy was more than happy to pay it. The crew carefully unpacked the Mark II and lenses and began bolting the camera to a projecting crane. Space at the lip was tight, so someone would gingerly balance each case on the nearby slope as it was emptied, wedging it against whatever boulder or root was available.

Nancy anchored into a safety line, leaning back to look at the river and cave below. As she did so, a crew member turned from one of the empty cases he had just set aside, a metal box that had made the journey to Everest and to the caves of Greenland and the Yucatán. It slipped forward a foot and was gone, tumbling silently through the

Dissolved calcium carbonate emerging from powerful cave springs colors the Little Colorado a milky white. Legendary cave diver Sheck Exley briefly entered these springs in the 1970s to find a cave more extensive than he could explore at the time; no one has entered them since.

desert air before slamming into rock with an echo that reverberated through the canyon like a volley of rifles.

In May 1869, Major John Wesley Powell, a war hero who had lost his right arm at Shiloh, rode the Union Pacific Railroad to its intersection with the watershed of the Colorado River at Green River Station, Wyoming. Under his command were ten men, charged with making the first descent through the Grand Canyon of the Colorado in order to survey the geography and geology of the largest parcel of still unexplored land within the United States. In four modified wooden dories, they were to make the descent of a river known to run wild with rapids and falls. Powell had funded the expedition with the combination of a small grant from Congress, some leftover War Department rations, and a donation by the Smithsonian Institution. One crew mem-

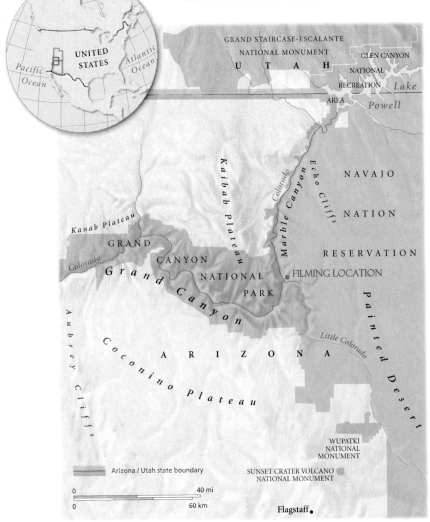

ber quit almost immediately upon realizing the dangers the trip would entail. Three others abandoned the effort halfway through, only to be killed by Shivwits Paiute Indians as they made their way out of the canyon.

Thirteen weeks after Powell set off, he emerged with six men in two boats and one of the great adventure stories of the 19th century. The successful descent helped him secure funding for a second, more scientifically qualified expedition in 1871. He combined his diaries of these two trips into a narrative that appeared in *Scribner's* magazine in 1874-75 and was later published in book form by the Smithsonian. The story's immediate popularity helped Powell become the first director of the United States Geological Survey and a founding member of the National Geographic Society. The book remains a classic of adventure and exploration. Yet almost from the moment it appeared, Powell's account has been a subject of controversy among historians and journalists.

In the interest of telling a fine story, Powell telescoped events, making no distinction between expeditions that took place two years apart by very different teams. The diaries consist of daily entries from the summer of 1869. According to those who were with him, many of those describe events of 1871, and even events from the first trip are often entered under the wrong date. While Powell relates particular discoveries and adventurous mishaps accurately, he often moves the locus of these events miles up or down the river as best suits the story. In other cases he changes the names of participants or the boats they used. He even includes his 1872 discovery of what eventually became Zion National Park as though it were a side excursion of the 1869 trip. Throughout the narrative, Powell takes dramatic liberties of the sort for which modern critics commonly castigate the creators of popular nonfiction books and documentary films.

Yet Powell's achievements remain undeniable historical fact, just as the book makes exciting reading today. By fictionalizing elements of the truth into a well-told story that grabbed the nation's imagination, Powell helped launch a century of government-sponsored exploration and wilderness preservation. Although they were few and incidental, Powell's observations of various caves and caverns within the canyon were often wholly accurate as well as poetic. These accounts likewise strayed from facts of time and location. In this, Powell may have anticipated a model of dissembling rigorously followed by the cavers who explore the Grand Canyon today.

The great caves of the Grand Canyon remain, by and large, closely guarded secrets, their glories occasionally whispered around late-night campfires at caving conventions, only to be denied the next morn-

ing. A motto of long standing, displayed on patches and T-shirts, boasts, "Arizona has no caves to speak of." The Arizona cavers don't speak of them, which may be why so many of the state's caverns remain in pristine condition, unaffected by the heavy human traffic that mars many Eastern caves. When Arizona cavers do publish photographs or articles, they are most often seeking government protection for a site or relating an important scientific find—and names and locations are intentionally obscured even in these documented cases.

An example is the recently opened Kartchner Caverns State Park near Tucson. The four Arizona cavers who discovered the large, well-decorated cavern kept it secret from other cavers and the public for 18 years. Throughout this time, they slowly persuaded state officials to acquire the land

Nancy pauses at the lip in 112° F heat before rappelling 350 feet to the cave entrance in the cliff below. Director Steve Judson called this narrow blade of sun-blasted rock "the scariest, most exposed location" he could recall in a long career spent making IMAX theater films in difficult places.

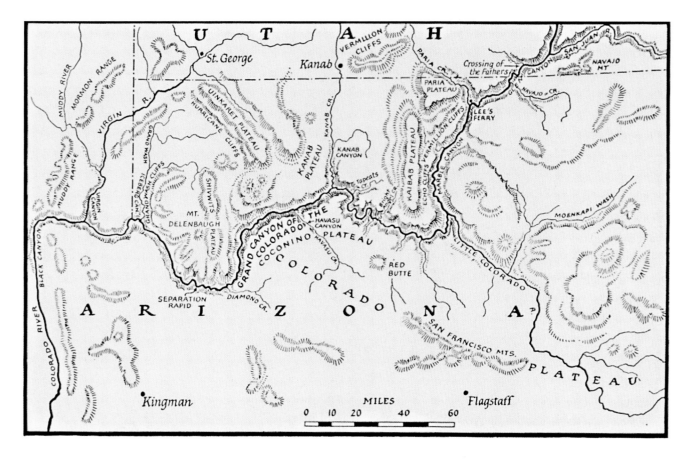

and to make efforts to protect the cave. They did this by showing slides of its impressive calcite columns to legislative committees, even by taking the state's governor underground—all without revealing the exact location to any but a few well-chosen allies, sworn to silence. When pressed on how to find the cave, the explorers would give vague or even false coordinates.

Meanwhile, they worked with conservationists and engineers to develop a tourism plan unlike that of any other public cavern. Trails and lighting would be laid discretely, without disturbing virgin cave floors inches away. The numbers of tourists allowed in would be strictly limited, to control the lint and other inadvertent human detritus that has affected such caves as Carlsbad and Mammoth. The end result of the long secrecy is that tours of Kartchner are now booked months and even years in advance. Each visitor sees the gleaming, stalagmite-lined passages exactly as the original discovers did.

Many other secrets remain hidden in Arizona limestone.

A few miles away from the cave in the Redwall limestone of the Little Colorado, four Navajo men of "The Big Rez" raised their eyes at the sound of the crashing lens case. Permission had not been granted lightly. These snaking canyons were hallowed ground to the Navajo, as to the neighboring Ute, Hopi, and Havasupi who had shared them since long before Powell descended them. The filmmakers had promised to use utmost care at the cliff, to leave no trace of their passing. Any accident, and the attendant rescue effort any accident would entail, might curse the ground, necessitating a complex cleansing ceremony to restore the balance of things.

After staring for a moment, eyes shaded against the intense sun, one of the elders pronounced that the sound was different from that a falling body would make, and therefore probably nothing to worry about. If the movie people had dropped something, they would pick it up, which they did: a volunteer eventually rappelled 500 feet to pluck the pancaked remains of the case from the limestone.

Left: "The river is rough, and bad rapids, in close succession, are found," Powell wrote in his journal for July 21, 1869. The engraving "Running a Rapid" left, published with Powell's account in Scribner's Monthly, was based on one of the many sketches of the second Grand Canyon expedition made by the talented young artist Frederick Dellenbaugh. Above: The map published with Powell's account shows the confluence of the Colorado and Little Colorado, due north of Flagstaff. According to Dellenbaugh, Powell's team camped on a sand bank in the Little Colorado not far below the IMAX filming cave.

The men returned to their work, preparing a site for the sand painting they agreed to create for the cameras the next day.

An artist who traveled with Powell's second expedition was Frederick Dellenbaugh. Only 17 at the time, his life as an explorer and naturalist was profoundly affected by the experience. Nearly 40 years after the 1873 trip, Dellenbaugh wrote *A Canyon Voyage: The Narrative of the Second Powell Expedition*. Unlike the accounts of Powell himself, Dellenbaugh's narrative scrupulously follows the diaries of team members in placing the time and location of events that happened along the way, and it often goes into far more detail on daily occurrences. One memorable passage describes the team's first encounter with the Navajo:

> We heard a sudden shout, and saw an Indian standing on the rocks not far away. We beckoned for him to come, and thereupon he fell back to another, and together they approached. We saw by their dress, so different from the Ute (red turbans, loose unbleached cotton shirts, native woven slashes at the waist, wide unbleached cotton trousers reaching to a little below the knee and there sashed up on the outer side for seven or eight inches, bright woven garters twisted around their red buckskin leggings below the knee, and red moccasins with turned up soles and silver buttons) that they were Navajo. They indicated that they were father and son, the father announcing himself in a lordly way as "Agua Grande." He was over six feet tall and apparently 60 or 70 years old. The son was a fine young lad of about 15. Their bearing was cordial, yet proud and dignified.

Several other Navajo arrived, and began exchanging speeches of welcome with the exploration party. Aqua Grande expressed concern over an obviously ill member of the expedition, and offered to transport him to the nearest white settlement, but the man refused to go. After a brief visit, the Navajos gathered on the banks of the Colorado to see the team off. "Clem, with his customary urbanity, went down the line all smiles, shaking each one cordially by the hand, and requesting him to 'Give all my love to all the folks at home,' and 'Remember me, please, to Eliza Jane,' and similar expressions," Dellenbaugh writes. "The Navajos did not understand the words, but being themselves great jokers they saw it was all in fun, and they all laughed, making remarks which doubtless were of the same kind."

The Navajo that were now clustered around the bright sand painting were dressed much as their ancestors had dressed in Dellenbaugh's day. Just as the only way to get accurate directions to a wild Arizona cave is to be an Arizona caver, the only way to see an authentic Navajo sand painting is to be an authentic Navajo. The approximations one sees in films, museums, and roadside attractions are just that: the images may look authentic, but any medicine man privy to the secrets of the sand painting ritual would not willingly rob the ceremony of its power. To reveal the exact symbols and beings depicted in a particular sand painting to an outsider would, according to Navajo beliefs, render the healing power of the painting moot. The spiritual result would be akin to effects of over prescription of modern antibiotics, making strong medicine suddenly ineffective.

To avoid such loss of power, the Navajo who demonstrate sand painting for others purposely leave out key details. They intentionally combine elements from different paintings that would never appear together in a genuine ceremony—sometimes in ways that would be as comically ridiculous in context as the line "Remember me, please, to Eliza Jane" was when spoken to old Agua Grande. Their audiences don't know or mind the differences, and their peers can tell at a glance that the painters have done nothing to profane sacred rites.

There are some cavers who approach the Grand Canyon with an almost shamanistic

A caver stands dwarfed beside massive columns deep within Lechuguilla Cave. Although the cave's main passage was first entered only in 1986, by 2000 more than 100 miles of highly decorated virgin tunnels had been mapped there. Lechuguilla has become a proving ground for microbiologists studying ecosystems wholly removed from sunlight, dependent entirely on mineral food sources.

reverence. And when in their presence, one senses the pride of the initiate. A combination of the secrecy and the extreme remoteness of the canyon's caves can lead some explorers to invoke their own rituals when seeking out new discoveries. Cavers dream of breaking into bejeweled cathedrals beneath mountains of limestone, places virgin still to light and humanity. Sometimes, with extreme effort, we find our dreams.

Donald Davis is a respected Colorado caver who has done a great deal of surveying in the Grand Canyon over the past four decades. He is also the author of a number of significant papers concerning cave geology and mineralogy. In 1982, he published a masterpiece of spurious science entitled "The Dilation Theory of Cavern Development." Davis and the paper's many fans insist the work is little more than a sophisticated inside joke. Even though he wrote it in jest, it has nonetheless been reprinted in several American caving periodicals and translated for German, Swiss, and Norwegian caving audiences. It has become the most far-reaching work of his distinguished (and otherwise legitimate) speleological career. Davis begins by proposing "that all previous investigators have been grossly in error in treating caves as geologic or hydrologic phenomena." Instead, he suggests, "the mechanism of cave creation is parapsychological, or more explicitly, psychokinetic." He continues:

> My own recognition of this fact occurred during a trip with two geologist-cavers in the Grand Canyon. They were mapping a measured section through the Redwall limestone cliff along a steep fault ravine, while I was leading, deliberately examining the walls for caves. None of us saw any caves during the descent. On returning up the route, we again saw no caves until the last man suddenly saw to his left, 20 feet away and in plain sight, an entrance ten feet square behind a Douglas-fir trunk. It led to a cave with 2,000 feet of passage. It had obviously dilated, or blinked into existence, at that final moment when the mental force of our party's attention to the matter had built up to the intensity necessary to create the cave. Cave dilation evidently operates according to quantum-mechanical laws in which the cave flips instantly from "off" to "on" when the threshold energy input has been reached, with no intermediate state. This is why no one sees caves opening slowly as they watch. Since it is well known that psychokinetic forces operate independently of the inverse-square law that governs normal physical interactions, most caves will be dilated at distances out of sight of the creators, and the caves will appear to have been there for much longer than they really have.

After providing an example of a Colorado cave unquestionably created through dilation, the author proposes an operative mechanism:

> Most caves are found in limestone, dolomite and gypsum because these rocks are associated with barren and unproductive landscapes whose inhabitants had little to entertain themselves but to relieve their boredom unconsciously by creating cavities underground. Once begun, the trend was self-accelerating by suggestion to others. When scientists then developed a wrong but superficially plausible explanation that water action caused the cavities, then caves dilated later (or even earlier, since psychokinesis may incorporate precognition) would tend to take on the features expected—a self-fulfilling prophecy.

The text continues in this wonderful vein, concluding with "Cautionary Notes" regarding the potential of widespread dilation research to precipitate black holes, by creating cave space that exceeds the volume of the planet.

Anyone who has done much caving recognizes the experience Davis describes—the cave that appears in plain sight only at the end of a long day spent looking for it, during which you passed the "obvious" entrance several times. There is an oft-proven corollary for major expeditions to large caves in foreign lands: on the last day of a big trip, as unwashed explorers contemplate the return to jobs, spouses, running water, etc., someone in the team will find a new passage leading in an unexpected direction, an immense borehole inevitably broken after a short distance by a vertical drop. This pit will prove deeper than the longest rope available, insuring that the new passage will remain unexplored until some future expedition. Such a "last-day pit" is virtually assured when there have been no other major finds during the course of a lengthy and difficult expedition, especially while in equatorial jungles.

Perhaps the most significant Western cave explored recent decades is Lechuguilla, located in Carlsbad Caverns National Park in southern New Mexico. The cave has grown from a known length of a few hundred feet in 1986 to over 100 miles today, and grows longer with every expedition. It is an ideal candidate for having been dilated into existence, not in the least because Donald Davis has been one of its principal explorers. Those who have been there with Donald say that he will sometimes enter a newfound chamber and pronounce in a demanding tone, "dilate spaciously"—the cave often complies.

Lechuguilla's single natural entrance was first observed in 1914 by John Ogle, a miner of bat guano who undoubtedly found himself inhabiting the barren and unproductive landscape with little to entertain himself. A shaft about 70 feet deep led to a pile of rocks from which a strong wind issued. In the early 1980s, a group of Colorado cavers became convinced that the source of the breeze was some vast unexplored cavern, and set about digging their way into it.

Over the course of several years, assorted small groups spent weekends poking at the rubble, and in 1986 a particularly determined party of four punched through to unknown miles of cave passage decorated by profuse and unusual formations. Because of the meandering nature of the cave tunnels, along with

unusually high concentrations of gypsum and other minerals, Donald Davis and other speleologists concluded that Lechuguilla had been formed not by surface runoff, as was usually the case with limestone caves. Instead, they posited that the cave appeared to have been carved from the bottom up, by rising springs rich in hydrogen sulfide, which could react chemically to form sulfuric acid and dissolve rock at a rapid rate.

In the early 1990s, this theory attracted microbiologists to the cave. Other deep, dark environments rich in hydrogen-sulfide, such as mid-ocean volcanic vents, had proven to contain treasure troves of odd bacteria that drew energy from chemicals in rocks. The life in these extreme environments was built around a chemical food chain—as opposed to the more familiar food chain of the surface world, based on energy gathered from sunlight. The cave contained thousands of species of previously uncataloged microorganisms living in complex communities. These ecosystems attracted microbiologists as well as NASA investigators to Lechuguilla, astrobiologists who hoped to find support for their expectations that primitive life existed beneath the icy surface of Mars and other planets. Inevitably, the new cave, the new bugs, and the NASA interest in turn attracted journalists. In 1996 I spent four days camped underground in Lechuguilla's nether reaches, assisting microbiologists in collecting rock-eating bacteria. Within the next two years, I collected microbial samples that were subsequently reported in NASA scientific papers that outlined principles for discovering fossilized microbes on Mars. I had become as much amateur scientist as reporter.

On my first trip out to the cave, I had met medical microbiologist Larry Mallory, who had done more work with novel cave microorganisms than any other researcher.

Branching, faceted selenite crystals up to 20 feet long festoon the Chandelier Ballroom of Lechuguilla Cave. Yet theirs is a delicate beauty: Should the chamber be opened to the public, geologists predict that the changes in air patterns and moisture caused by frequent traffic might send them crashing to the floor.

"If I wanted to collect a redwood tree, a tiger, and a trout," Mallory had explained to me, "I'd be collecting much more closely related organisms than many that live in a single water droplet hanging off a single helictite in Lechuguilla. There's an incredible diversity underground. In a given pool, you might find one community that lives in the film on the surface, another on the edge of the water, another on the bottom, and maybe another that stays suspended in the middle. Each of these communities might have dozens or even hundreds of separate taxa. They've been evolving and adapting for a very long time and are very different from one another. That's one reason they're hard to collect: You need several different techniques for each site, or you're going to miss most of what's there."

I asked whether any of these weird bugs could hurt me. Might some "Andromeda strain" be down there waiting for us to stumble onto it?

"Not likely," said Larry. "Parasitic relationships require a close shared evolution between host and parasite. Only familiarity can breed disease."

The twin ideas that unknown kingdoms of life lay waiting to be found, and that, despite my lack of scientific training, I might help find them, had proved irresistible. Tradition in American journalism holds that the reporter must remain detached from his subject, a mere observer as opposed to participant. Yet there has always been a contrary undercurrent of "participatory journalism," in which one does something exciting and writes about it. Such is the source of narrative power in the works of Peter Freuchen, John Lloyd Stephens, and John Wesley Powell. Moreover, quantum theory, I realized—or perhaps I should say rationalized—teaches us that the act of observation itself profoundly affects the observed phenomenon, to the point of determining its very reality. If this is so,

journalistic objectivity becomes illusory at best.

Like John Wesley Powell and Donald Davis, like the Navajo who sat at cliffside amid bowls of colored sand and the cavers and filmmakers who dangled precariously from a metal crane, I wanted to discover beauty in the unknown, to create art that helped define reality. A few weeks before Hazel and Nancy had come to Arizona to shoot the final scenes, each had agreed to take me to her favorite wild cave in order to

experience those elements of cave exploration and science that could not fit into a 40-minute documentary. Without camera crews or scripted agendas forbidding actual exploration, I set off with Nancy in search of virgin passageways and then with Hazel in search of new forms of underground life. I cannot relate the outcome of these searches—nor what lay beyond the virgin opening in the wall of the canyon of the Little Colorado River—without committing small inaccuracies in order to keep certain promises and to best tell the story.

But the essential truth of the following pages remains: This time the magic worked. The caves dilated spaciously.

Left: A kayaker in the Little Colorado shoots over a travertine dam, the surface equivalent of a cave formation. Microbiologists have found that travertine is invariably riddled with microbial fossils. Some believe the microbes deposit the calcite mineral, much as larger organisms build coral reefs. Above: a Navajo family practices the ancient healing ritual of sand painting.

CHAPTER 10

IN THE BELLY
OF THE PLANET

On a warm Saturday evening near the end of March, I arrived at a comfortable log house located at the end of a gravel road off a winding state highway in northeast Alabama. With me was Nancy, her husband Brent, and two mutual friends from Atlanta, Alan and Benjii. We were going, the next day, to survey in State Line Cave, one of many on Fox Mountain, an area of special interest to the Aulenbachs.

Above: A salamander deep inside Cagle's Chasm, a complex TAG cave, will exit nightly to feed. Right: A caver drops into a 90-foot "free" rappel in Thunder Hole.

That night, we arranged to sleep at the home of cavers Jim and Gail Wilbanks. ■ On any given weekend the Wilbankses' house fills with friends and strangers. People they know well and others just met converse on the deck or in the spacious homemade hot tub until well after midnight. They pitch tents or crawl into the backs of pickup trucks with camper tops and homemade sleeping lofts—a popular setup among TAG cavers. The couple never wants for company because they live smack in the middle of TAG, surrounded by deep pits and "going" cave systems (that is, caves containing unexplored leads), and because like many TAG cavers they possess generous hearts. ■ In addition to providing free lodging for cavers from around the country, they volunteer to help remove trash from polluted caves and construct cave gates at the request of landowners worried about unauthorized visitors. They train novices and organize cave conventions.

The Wilbankses donate a large portion of their time and income to the Southeastern Cave Conservancy, a nonprofit organization that purchases land surrounding TAG caves and pits in order to preserve them for future generations of cavers and scientists; Jim sits on the board. Both are active in local cave clubs as well as regional and national caving associations.

Another group of Atlanta cavers, who were targeting a different cave for the weekend, had already arrived at the Wilbankses'. I recognized one as Jerry Wallace, a cartoonist who had been illustrating caving magazines for decades. (I used to have a T-shirt from his "Great Lies of Caving" series: "Lie Number 3: You don't need knee pads" depicts a wincing caver crawling over hundreds of rocky spikes). Jim mentioned that several more carloads of students from a Pennsylvania outing club were due to arrive later that night and set up camp out back.

When Nancy had described Jim Wilbanks, I knew he sounded familiar, but I couldn't picture him. Once in his house, however, I immediately recognized him as a voluble caver with whom I had conversed often at previous conventions. What surprised me even more than knowing Jim was the sudden realization that I had stayed at his place once before, camping in the yard with some Florida friends who had brought me there in the 1980s, when the house was still under construction. Amid a friendly atmosphere that at times felt like a family reunion, we talked about caves and cave politics.

Nancy had asked me not to bring up the film. Like many cavers, Jim was nervous about publicity. Although he and the Aulenbachs were good friends, he had tried to talk Nancy out of appearing in the film. Too often in the past television documentaries produced in TAG had portrayed caving as an adventure sport for the adrenaline-addled. Such programs seldom mentioned the strong conservation and safety ethics of most cavers, let alone the science, cartography, landowner relations, and other dull but vital tasks to which TAG cavers devoted great time and energy.

Jim and others feared that the film—any film—would present little more than an extended Mountain Dew commercial, that it could entice untrained teenagers to harm caves or themselves as they tried to re-create stunts they saw onscreen. The end result would be more caves on private land closed to exploration, and more caves on public land suffering from increased visitation. Nancy and other cavers who supported *Journey Into Amazing Caves* argued that public education was the only path to widespread public support for cave protection. Jim had come down firmly in the opposing camp. Rather than launch into an argument that would carry long into the night with no resolution, we avoided the topic altogether, focusing instead on the caving goals of the next day.

Brent unfurled a scroll of paper three feet wide and eight long, draping it over the edges of the coffee table in the Wilbankses' living room. It was a map in progress of State Line Cave, showing the major passages explored to date. The tortuous route from the entrance intersected a major stream passage—not yet surveyed, but roughly penciled in—snaking northeast to southwest for nearly a mile beneath Fox Mountain. The stream crossed the Georgia-Alabama state line along its route. "When we go in tomorrow, we won't have any problem with the landowner where we park, or the landowner of the cave," explained Brent. "But right here"—he pointed to a triangle penciled on the topographic map—"we have to walk along the edge of the field of a farmer who doesn't much want cavers on his property. He really can't keep us out because we have permission of the cave owner, but we try to stay to the side of his property and cross as quietly and quickly as we can."

"Actually, thanks to the SCCI we have wonderful relations with most of the landowners up there," Jim explained. "The old survey records were confused as to where the property lines were, so we volunteered to resurvey their property lines for them. Brent and Nancy did some of the work." Jim had organized 75 volunteers to survey a 233-acre cave preserve.

Nancy, top, takes the high road and her husband Brent takes the low as the couple explores a flowstone-choked passage in TAG. One of them has briefly set aside a nylon cave pack on the ledge at left while they check every lead, no matter how tight.

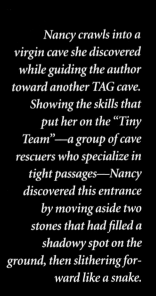

Nancy crawls into a virgin cave she discovered while guiding the author toward another TAG cave. Showing the skills that put her on the "Tiny Team"—a group of cave rescuers who specialize in tight passages—Nancy discovered this entrance by moving aside two stones that had filled a shadowy spot on the ground, then slithering forward like a snake.

In several cases, key information came from neighbors to the preserve, who were grateful for an accurate survey at no cost to them.

In the process of respecting the rural landowners, one of them shared the location of some cave entrances that had never been explored. The often difficult process of obtaining permission to explore caves on private property is one reason so many cavers shy away from publicity. The more people crossing a farmer's fields, the greater the farmer's fear of open gates and lost stock, trampled crops, or—even worse—of some careless spelunker getting hurt and filing a lawsuit. To alleviate such fears, caving lawyers have helped develop liability release forms for landowners. Local cavers like the Wilbankses keep them on hand and make sure anyone heading onto a particular piece of property signs the proper form. In a few rare instances releases have been challenged in court, but those that are issued by TAG cavers have successfully absolved landowners of all responsibility for the safety of explorers entering their caves. Jim was in the process of setting up a website called "The Cave Access Project," which would provide accurate cave liability information for all southeastern states, so that landowners and cavers alike would know the

local liability laws when discussing permission to explore a given cave. All the same, the lower the foot traffic to most caves, the better the relationship with the owners.

"I think the main stream passage ties hydrologically to several springs down here," Brent said, pointing to a marshy area on the edge of Fox Mountain, where he said he planned to do a dye trace. But the stream disappears both upstream and down"—he pointed out the ends of exploration on the map—"so the only hope of seeing more of it is to find a bypass. Right now, that seems unlikely, but there's quite a bit of mapping left to do, and there's at least a few leads left."

"And on the way to the stream passage, you'll get to see some real fun cave," Nancy added with a smile, her eyes twinkling.

The next day dawned cool and cloudless. After loading up at a local diner, we drove to a remote farmhouse and parked alongside the road. The residents were out, so we left a detailed note listing the number in our party and when we expected to exit. Each of us signed a release form, and Brent left all the paperwork in the owner's mailbox. Then we set out off through fields and woods, following streams and a few crumbling fence lines up the mountain. We stopped at the ruins of a settler's cabin—two walls of sturdy limestone blocks covered with moss. I spotted a black snake between four and five feet long stretched out on a warm boulder, a cloudy patch of loose skin peeling from head to reveal a shiny new layer beneath. Alan picked it up to take its picture.

"That's a black racer," he said, handing it to Nancy to hold. "These guys are incredibly fast. If he weren't molting, he'd have been gone before we got near him." Nancy returned him gently to his rock. Here the path to the cave turned upward, following a streambed through an exposed limestone ridge. From long habit, we spread out from each other as we worked our way over the ridge, eyes scanning the shadows for hints of an entrance. New caves have been discovered in TAG only a few

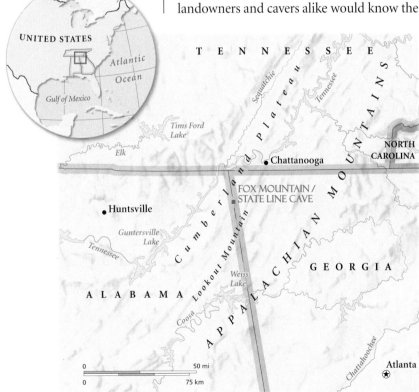

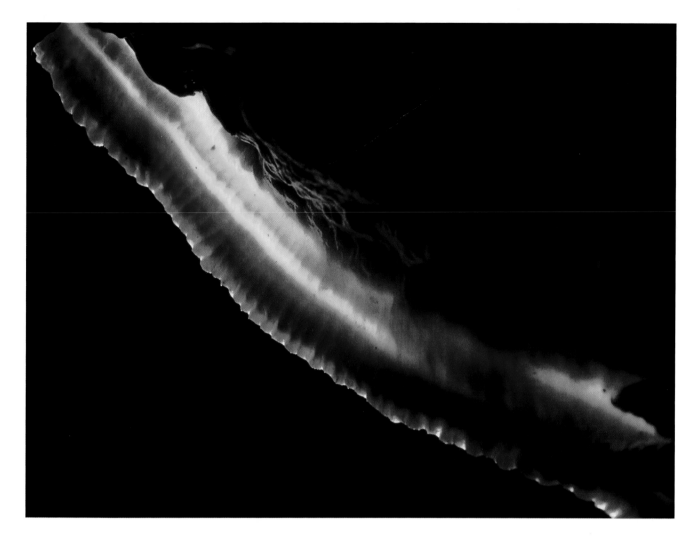

feet off well-traveled trails leading to other caves that had been mapped decades earlier.

This was a perfect day and location for "ridge walking," the process by which new caves are found. The spring growth had not yet filled out the underbrush to the point where it would block all on the mountain from view. The contact between limestone and sandstone stood out plainly, marked by a "ridge" of limestone boulders and outcrops along which one could "walk" in search of small holes and openings. With the geology of Fox Mountain, any shadowy hole along the ridge line might lead to an extensive cave or pit. And the cool spring weather was ideal for hiking along a mountainside.

At the same instant Nancy and I both spotted what looked like the mouth of a small cave, only a few inches wide. It lay at the base of a short limestone face. Nancy stepped up to it and began pulling aside loose stones, quickly enlarging the hole to a diameter of about two feet.

"Hey, Brent," she called. "You ever see this before?"

We waited a few minutes while Brent and the others hiked over. "That's definitely new to me," he said. "Check it out."

He had barely finished speaking before Nancy had vanished through the small opening—the cave was at least big enough for one. "I see passage," came Nancy's muffled voice from inside. Brent, Alan, and Andy all disappeared inside. I could see their lights no more than 20 feet ahead, and I could see that at least one of them had room to stand. I lay on the moss and dead leaves to follow. The entrance was tight; I had to remove my helmet to keep my lamp from snagging. I shoved it ahead of me and wriggled past small spines of rock that scraped against my chest, then crawled down a short slope into an irregularly shaped chamber, its center just high enough for two or three people to stand.

Bands of iron deposits give "cave bacon" its characteristic stripes of color, best seen by holding a flashlight or helmet lamp behind the translucent formation.

KNOWING WHERE YOU'VE BEEN

BY MICHAEL RAY TAYLOR

Ariadne instructs Theseus to carry a ball of thread with him to the labyrinth of the dreaded Minotaur. He ties one end of the thread to a door lintel, unrolling the ball as he makes his way through the maze. The hero kills the monster and follows the thread back to daylight.

Greek mythology's advice on navigation falls short. Literally. Some casual weekend spelunkers still use Theseus's method through complex cave routes, and quickly realize that their longest ball of twine soon runs out in even a relatively short cave. Entrances of well-known, well-traveled caves are often littered with lengths of string. Some novices are even guilty of defacing cave intersections with spray-painted arrows labeled (often incorrectly) "This Way Out." All this leaves for organized, experienced cavers the task of cleaning up the sites—collecting abandoned string and scheduling, as many cave clubs do, annual cleanup days to scrub away unnecessary and misleading graffiti.

Rather than resorting to twine, spray paint, bread crumbs, or the like, experienced cavers can avoid becoming lost in complex, multilevel cave systems simply by developing their powers of observation. Since irregular tunnels seldom look the same way going out as they did coming in, a caver learns to constantly look over her shoulder, memorizing passage details such as formations and rock striations. At each intersection, a care-

ful caver will pause and note which passage she is leaving, which she is heading into, and any other possible routes not chosen. Just as seasoned wilderness hikers develop an innate sense of where they are in relation to major rivers, mountains, or trails, cavers become adept at relating their location to major landmarks within the cave. No matter how labyrinthine the crawlways, cavers seldom get lost. The more honest among us, however, can relate to Daniel Boone, who claimed never to have been lost in the woods, but once admitted to being "bewildered for three days." The trick is to carefully retrace one's steps to a recognized spot—an approach perhaps familiar to those of us who have "gotten lost" in today's enormous shopping malls and their equally disorienting parking lots.

While skill and observation are essential for cavers in order to return safely to the surface, accurately describing where they've been is also essential, and requires the ability to create maps. It is extraordinarily difficult to estimate lengths underground. A caver might walk through a mile of easy stream passage in less time than it takes to squeeze through 100 feet of difficult crawlway. Passages seldom run straight, and intersections seldom occur at right angles: it is possible to spend hours circling around and around just below the entrance. The best way to begin to find new passages within a known cave is to understand its geologic nature. Does it follow a particular fault or ridge? Is it

related to surface topography? Does an ancient drainage pattern govern the major passageways?

These are questions that can be answered only with a carefully drawn map, and any caving expedition that fails to produce maps is often regarded as a failure. While cavers may scout a short distance down a new cave or passage in order to determine whether the find is worth mapping, caving etiquette requires all significant discoveries to be mapped as they are first explored. It is only through a careful plotting of where cavers have been that a complex system of caves can be studied. The making of cave maps is thus an art that requires both technical survey proficiency and the talent to sketch. Every year, the National Speleological Society awards prizes to the best maps produced on national and international expeditions.

Cave cartography depends on three instruments: a measuring tape, a compass, and a clinometer—clearly, a device that measures inclination. Cavers set survey stations—usually marked by a bit of plastic flagging—throughout the cave, measuring distance, compass bearing, and degree of inclination from each to each. By plotting this recorded data on a computer program—connecting the dots from station to station—a simple but accurate line map of the cave is created. Fleshing this out into a true representation of the cave depends on the skill of the sketcher, who makes minute

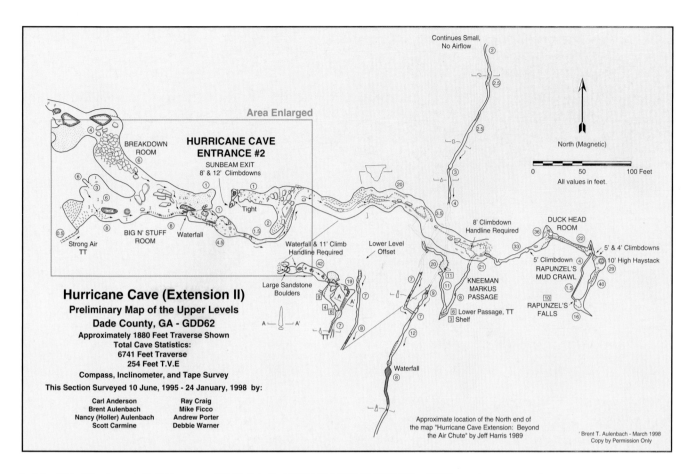

Area Enlarged

Continues Small, No Airflow

HURRICANE CAVE ENTRANCE #2

SUNBEAM EXIT
8' & 12' Climbdowns

North (Magnetic)

0   50   100 Feet
All values in feet.

BREAKDOWN ROOM

Tight

BIG N' STUFF ROOM   Waterfall

Strong Air TT

8' Climbdown Handline Required

DUCK HEAD ROOM

5' & 4' Climbdowns

5' Climbdown   10' High Haystack
RAPUNZEL'S MUD CRAWL

Waterfall & 11' Climb Handline Required

Lower Level Offset

RAPUNZEL'S FALLS

Large Sandstone Boulders

KNEEMAN MARKUS PASSAGE

Lower Passage, TT
Shelf

Hurricane Cave (Extension II)
Preliminary Map of the Upper Levels
Dade County, GA - GDD62
Approximately 1880 Feet Traverse Shown
Total Cave Statistics:
6741 Feet Traverse
254 Feet T.V.E
Compass, Inclinometer, and Tape Survey
This Section Surveyed 10 June, 1995 - 24 January, 1998 by:

Carl Anderson	Ray Craig
Brent Aulenbach	Mike Ficco
Nancy (Holler) Aulenbach	Andrew Porter
Scott Carmine	Debbie Warner

Waterfall

Approximate location of the North end of
the map "Hurricane Cave Extension: Beyond
the Air Chute" by Jeff Harris 1989

' Brent T. Aulenbach - March 1998
Copy by Permission Only

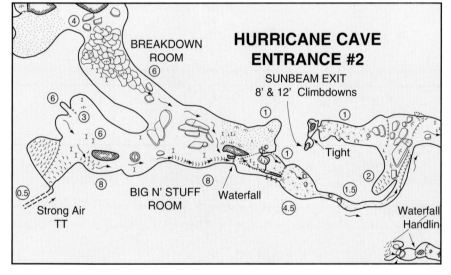

BREAKDOWN ROOM

HURRICANE CAVE ENTRANCE #2

SUNBEAM EXIT
8' & 12' Climbdowns

Tight

Strong Air TT

BIG N' STUFF ROOM   Waterfall

Waterfall
Handlin

The twisting of passages can disorient even experienced cavers. Careful cartography depicts a cave's streams and passage trends, giving clues to the cave's formation and where undiscovered passages might lie.

Springs, a complex underwater cave system in north Florida. Using scooters equipped with side-scanning radar and underwater radio links to mainframe computers. Stone's crew scanned the cave system by swimming through it, then took the millions of data points to create maps at a level of exactness never before achieved in the mapping of any natural structure.

While cartographers like Hazel Barton are not giving up their drawing pens any time soon, such electronic mapping devices appear to be the wave of the future. The cavers who one day explore the lava tubes of Mars may have an instant map readout the moment they enter a passage and as they travel a newfound passage. Meanwhile, a caver must continue to navigate and chart the Earth's underground labyrinths by looking over her shoulder.

drawings of passage features between the labeled survey stations. These drawings and the numerical survey data guide the person who finally drafts the map in creating both plan and profile views of the cave.

Whether it is a simple stream crawl mapped in an afternoon or a 100-mile system mapped over several decades, the process of mapping can be slow and tedious. In underwater caves—where the breathable air on your back

is measured in minutes—cave divers have created shortcuts, streamlining the process followed in air-filled caverns. The result is that maps of underwater caves systems are widely regarded as less accurate than those of traditional caves. Working under grants from the National Geographic Society and other organizations, cave diver Bill Stone recently developed a technology to create the most accurate map ever rendered when he charted Wakulla

"Hey, it's not much, but it's definitely a cave," said Nancy. The crawlway continued on the other side of the entrance, but at even smaller dimensions.

"Any airflow?" I asked. A breeze was often the best indicator of more passage beyond.

"Maybe a little." Alan said he felt some air, too. But the crawl was decidedly smaller than Nancy, who was decidedly smaller than rest of us. She was part of an informal group of TAG cavers called "the tiny team," trained specifically in rescuing injured cavers from passages too small for larger rescuers to fit into. One of Nancy's great regrets concerning the film was that a planned expedition to TAG, where the large-format camera would be somehow wedged into a true horror crawl, had ultimately been scrapped as too costly and difficult to film.

"Hazel gets to give the audience claustrophobia in the underwater squeeze in the Yucatán," she had told me. "I wanted to show them TAG caves, with all their mud, tight squeezes, and vertical shafts and streams." Evidently, she planned to have me fill in for the audience.

But not in this new cave. If the hole was too tight for Nancy, it was too tight, period. It might be worth digging at someday, but for now we had seen all 50 feet or so of "easy" passage that this cave had to offer. Brent would add a new dot on the topo map, a new possibility for future work, and we would continue on up the mountain to our goal.

Properly baptized in mud, we left the little cave. The day was getting warmer. After another 20 minutes' hike, we approached the rim of a craterlike depression. As we climbed up the downhill side, I could tell by the sunken trees on the other side of the rim that the collapsed area could have held all of the cave we had just discovered, along with two or three barns and an average-size airplane hangar. A sinkhole such as this constituted a classic TAG advertisement for "going" cave: if there were not an outlet at the bottom, the sink would have become a deep lake with the first big rain. The sinkhole was dry; ergo it had to contain the entrance to a descending cave.

"Gee, you think there's any cave around here?" I asked.

"Nah," Nancy answered. "How could there be a cave here?"

We reached the edge of the sink and I was hit by a blast of cool air, the natural air-conditioning provided by a blowing cave. A small stream cascaded into the sink from farther up the mountain, bouncing over a series of fern-draped falls. Nearly a hundred feet below us, at the base of the mossy, cone-shaped sinkhole, the stream vanished into the low wide entrance to State Line Cave. The cool wind that reached us at the top of the sink carried the loamy smell I have always associated with caves.

Until I began accompanying microbiologists on collection trips, I hadn't wondered why nearly all caves I had been in smelled the same. The odor common to caves and moist garden soil is actually the smell of actinomycetes, a type of bacteria that grows in long, threadlike colonies. One day I was in the lab of microbiologist Larry Mallory when he opened a petri dish containing live cultures from Lechuguilla cave. He had collected the cultures by swiping a tiny metal tool across some cave soil and then placing a few grains of it in a specialized growth medium. Larry had left the dishes sealed in his lab for months, in a dark environment that mimicked the cave; slowly, yellow and red circles indicating microbial colonies had appeared. When Larry opened one of the dishes in his lab, the smell that instantly saturated the room was the same as that rising from the entrance before me now.

At the Wilbankses' the night before, Brent had explained the sink with its obvious entrance had in fact been explored in the 1988, but the original explorers found only a few hundred feet of passage. It was only the recent discovery of the Hog Wallow that led to the main part of the cave.

We had checked out the small entrance en route wearing lightweight hiking clothes, but now that we had reached the cave proper, we opened our packs and began pulling out polypropylene long underwear and heavy nylon coveralls. I had left my coveralls at home

Nancy splashes through a shallow pool in a Georgia cave she and her husband Brent have explored and mapped extensively. The heavily scalloped limestone provides evidence of the massive water flow that forms most TAG caves.

when exploring the warm caves of the Yucatán, but I knew I would need them in the wet cave ahead. They were made of the same sturdy material used for bulletproof vests, yet one trip to a cold, wet California cave the previous year had split several seams. The day before we left Atlanta, I had an hour in Brent and Nancy's garage repairing the damage, gluing fresh nylon patches over the holes with a wonder goop called Canvas Grip.

After falling and flooding, the greatest danger caves present to those who explore them is hypothermia; proper clothing is the best defense against cold that can sap strength and impair judgment. Contact with the damp cave air, mud, and especially cave streams and pools leaches away body heat. Cotton and other natural fibers that provide warmth when dry become useless when wet. Through trial and innovation, cavers have found combinations of strong artificial fibers that can keep an explorer warm for hours or even days underground. Several caving entrepreneurs manufacture specialized cave clothing as a cottage industry, making most of their sales at cave conventions. One reason the current generation often explores depths untouched by earlier cavers is that it is better dressed.

Once geared up, we began a long scramble downward, wiggling through and around large breakdown boulders. I could sometimes hear the stream chuckling nearby, but it followed a smaller path downward so we took the route Nancy and Brent had found. We climbed down a number of short chimneys and rock faces before the main passage began to level out in a wide, gravel-filled crawlway. In places the ceiling was tall enough that we could walk bent over or in a squatting duckwalk; in other places it was strictly hands and knees, and I was grateful for the kneepads beneath my coveralls as I scrambled along for several hundred feet. As in pro football, many a caving career has ended abruptly due to knee injuries.

The ceiling height gradually dropped, though, so we had to drop from hands and knees to our bellies. Then the gravel gave way to mud. At last we reached a low room, wide enough for several to lie in at the same time. On the far side a troughlike depression filled with four or five inches of muddy water snaked around a corner, clearly the only space large enough for a human to pass. Nancy, who had been leading, offered to let me crawl ahead.

"Welcome to the Hog Wallow," she said in a voice hinting at pride of ownership. Nancy was one of the original mappers of the place. "Just follow the water—I guarantee you won't get lost."

I turned my head sideways, to keep both my light and nose above water, and slid in with a splash, like an alligator entering its den. The bottom and sides were slick mud, over which I didn't so much as crawl but squirm, paddling forward with my hands and toes. There was a good seven or eight inches of airspace above the water—much more than in some crawls I had traversed—but my helmet still scraped the ceiling. My neck began to stiffen from holding it sideways, the best position to both breathe and avoid banging helmet and lamp. I knew from the map that I had to traverse the length of a couple of football fields in this fashion.

I recalled Number 4 of the "Great Lies of Caving" series: "It's not wet." That cartoon shows a caver's face pressed against the ceiling above a frothy stream, his comically extended nostril snatching air from an inch of space. I had been places like that, and luckily this wasn't one of them. In comparison to, say, the Grim Crawl of Death, a 1,000-foot horror crawl beneath the Big Horn Mountains of Wyoming, the Hog Wallow really wasn't wet. At least, not that wet. I was glad for my warm coveralls, and at the same time glad for the coolness of the water because squirming through the muddy wormhole required such exertion. The mud clung to my gloves and boots, letting go, reluctantly, with slurps and gurgles. The sounds, I imagined would be the same if I were to crawl through the digestive tract of a great beast.

After an interminable 30 minutes, I popped out at last in a slightly larger, gravel-lined passage resembling the one before the Hog Wallow. The flowing stream from the surface

rejoined this tunnel, and around the next bend I could suddenly stand. I understood why we had put up with the misery behind, already forgotten. Just ahead of me the stream shot out of the tunnel into a waterfall, crashing 40 feet below into the floor of a chamber that stretched far overhead. Black nodules of chert poked from water-polished walls of yellowish limestone. I stood at the lip looking over the room, grateful that Nancy had let me lead into this stately hall. I could see the shadowy entrances to other tunnels at several levels both above and below. This junction marked the beginning of the main, "going" portion of the cave. While most of the leads branching from it had been explored, none had yet been mapped, a condition we would begin to correct momentarily.

"That leads to the main stream passage," Nancy said as she emerged from the crawls. She pointed to a rounded arch across the room and below us. "There's some really nice passage in the downstream section. I expect we'll go there first."

My past few caving trips had taken place amid the assault-force attitude of large international expeditions. Now I was enormously glad to be part of a small team in TAG, exploring a lovely if not quite world-class cave on a Sunday afternoon. This was what caving was about: getting to know the land and the culture of a people and of a place, walking the ridges, meeting the residents, studying the topography and hydrology until you could point to a spot with authority and say, the cave must be here. And while "nice passage" on an

Nancy's husband Brent proposed on rope as the two climbed out of this "classic" TAG pit, called Neversink, located in northeast Alabama. The fern-draped 180-foot shaft is owned by the Southeastern Cave Conservancy, a nonprofit organization that preserves caves for future generations of cavers and scientists.

HOW CAVES FORM

By Arthur N. Palmer, Ph.D.

The most powerful cave-forming agent is underground water, which widens subsurface cracks as it flows through them. In glacial caves, the original cracks result from flexing of the ice as the glacier moves slowly under its own massive weight. Melting ice and snow during warm weather causes water to pour into the cracks, gouging deep vertical fissures as it melts their icy walls. Fissures are usually widest just below the surface because their tops are partly covered by windblown snow. They narrow with depth as the original cracks get tighter and the water temperature drops.

When a crack becomes too tight to carry all the water, the overflow follows horizontal cracks and forms lateral tunnels. In small glaciers the water makes it all the way to the bottom, moving along the base of the glacier and emerging at springs at lower elevations. In large glaciers, like those of Greenland, most of the water simply freezes far below the surface, sealing up the deepest cracks even while the higher ones continue to enlarge.

Much more permanent and far larger than ice caves are those formed when underground water dissolves bedrock. The only rocks soluble enough to allow such caves to form began as chemical deposits, usually on an ancient seafloor. Limestone is

the most abundant of these. It contains shell fragments cemented together by calcium carbonate from the seawater. These rocks have since been squeezed upward by forces within the earth, exposing them at

Minerals decorating caves typically grow after cave-creating processes are complete; they are called secondary formations.

the surface, where fresh water can pass through and form caves.

Cave-forming water begins as rain and snowmelt, seeping through the soil into cracks in the underlying rock. Water's ability to form caves is greatly increased by small amounts of acid. On its way through the soil, the

water picks up carbon dioxide from organic reactions, forming a mild solution of carbonic. Other acids can come from deep beneath the earth—for example, sulfuric acid is produced when hydrogen sulfide (a gas common in oil fields) rises and reacts with oxygen near the water table. This acid formed some of the caves in the Grand Canyon as well as the caves of Carlsbad Caverns National Park.

Where it first enters the ground, water descends by gravity. If it flows straight downward along cracks in limestone it eventually creates vertical shafts shaped like wells. Some plunge straight down for hundreds of feet. The deep pits in the southeastern United States formed in this way. Some are fissures shaped like those of Greenland. Where water follows gently sloping cracks, it dissolves their floors, forming narrow winding passages that look like stream canyons with roofs. The water eventually reaches a zone where all openings in the rock are filled.

The top of this zone is called the water table, which is approximately the level that water stands in wells. At and below the water table, water follows the paths of least resistance to springs in nearby valleys. It forms tube-shaped cave passages with rounded cross sections. Most of the Yucatán caves formed beneath the water table.

Only those cracks that carry the

most water eventually become caves. As caves enlarge, sinkholes and sinking streams form at the surface, through which most of the water enters the ground to feed the growing cave system below. Individual cave passages tend to join as tributaries, just as surface streams do. But unlike surface streams, caves can branch in three dimensions, forming complex levels and interconnections.

The carbon dioxide picked up by the water from the soil is more concentrated than in cave air. When water trickles into air-filled caves, it loses some carbon dioxide to the cave air. This water has become nearly saturated with dissolved limestone on its way to the cave, and when it loses carbon dioxide it is able to hold less limestone. Some limestone is deposited in the cave, creating crystalline formations such as stalactites and stalagmites. Cave pools acquire a lining of crystals, such as the dogtooth spar found in IMAX Cave in the Grand Canyon. In dry caves water can evaporate, depositing minerals as needlelike crystals, delicate cottony tufts, and flowerlike fronds. These beautiful mineral formations typically grow long after the processes that created the cave are completed—hence they are called secondary formations. Secondary formations such as those shown on this page are signs of mature cave passages, whose formation process ceased long ago.

Caves of the Yucatán formed at or below the water table, which is essentially at sea level, and they remain water-filled today; because of this, they are probably still growing. But they also spent considerable time in the past above the water table, as shown by their many stalactites and stalactites. A lot of water is tied up in

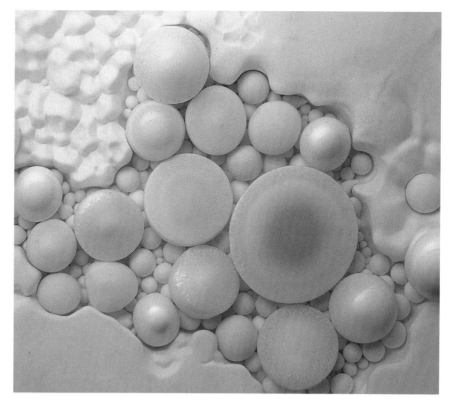

Above: Calcium carbonate "cave pearls" typically form in the bottom of small pools. Left: Crystals branching from dripping soda straw stalactite from an aragonite "bush" in Lechuguilla Cave.

glaciers; as climate changes cause them to grow and retreat, sea level drops and rises. The stalactites and stalagmites in Yucatán caves tell us of a time when glaciers were much larger than today and sea level was several hundred feet lower.

There are other ways to form a cave. Some small ones come from the weathering of rocks at the surface, where weak or fissured rocks crumble more rapidly than strong ones. Others occur when a volcano erupts and spills lava down its sides. As the surface of the lava solidifies, the still-liquid lava below flows out, leaving underground tunnels called lava tubes.

The lifetime of a limestone cave is usually a few million years. Shallow bedrock caves, including lava caves, last only a few thousand years. Eventually the eroding land surface intersects them and they are gradually destroyed. Glacier caves last only a few decades at most before they are sealed shut by glacial movement. But as old caves are destroyed, new ones grow to take their place, so the pattern of caves constantly changes throughout the life of the planet.

expedition to China or Borneo might be defined by stadium-size chambers and formations the size of rocket boosters, for most weekend cavers small beauties would suffice: a scalloped limestone wall that scattered a reflected shimmer of helmet lamp from the stream; a delicate gypsum flower hidden between boulders like a secret orchid of stone; a crawlway lined with bubbles of white popcorn formation, each no bigger than a pea. Small wonders were wonders all the same, all the more wonderful when shared only among the few cavers who had discovered them.

And then came the thrill of caving itself—nothing heroic or dangerously athletic, but intensely physical all the same, demanding a quiet competence that becomes its own reward. Perhaps it was for the best that the finished film would depict stunning vertical drops and vast underwater chambers, but none of the muddy, unremarkable crawlways in which cavers spend most of their time underground. The thrill of such places has to be experienced in the flesh.

Nancy pulled mapping gear from her pack and Alan began to rig a rope beside the waterfall. I looked out at the darkness and hooted, my voice echoing above the music of the falls. When John Wesley Powell had entered a particular "great hollow dome" in the Grand Canyon, I recalled, his team had done the same thing. "Standing opposite the rock," he wrote, "our words are repeated with starting clearness, but in a soft, mellow tone, that transforms them into magical music." This cave had its own magical song, and as we stretched the measuring tape from point to point, calling out distances and inclinations and compass readings, our words condensing in plumes of fog as though we were dragons, the cave sang to us.

I savored the song, trying not to think too much about the Hog Wallow we would have to traverse once more to reach the surface that night. I took comfort in the knowledge that the difficulty of the passage—and the dedication of TAG cavers like Nancy Aulenbach—would protect State Line Cave's glories for generations to come.

Nancy takes a compass reading between survey stations in a damp TAG cave. By recording compass bearing, degrees of inclination, and distance between stations in the cave, explorers can later construct an accurate map of passages on the surface.

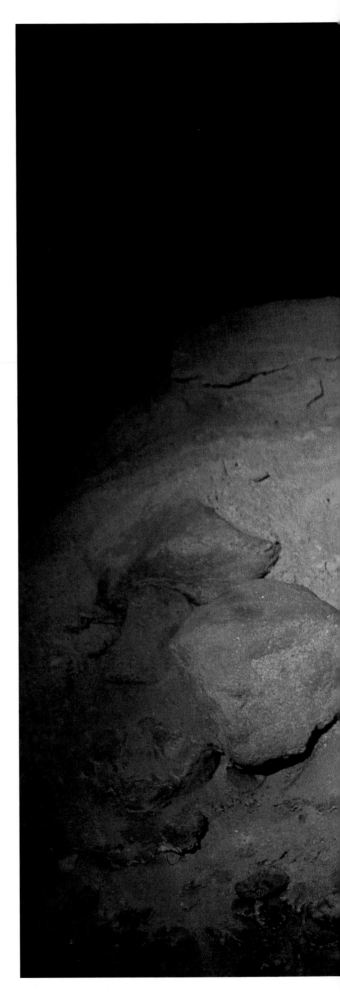

CHAPTER 11

DOES IT GO?

azel Barton drives a beat-up four-wheel-drive truck bearing an assortment of scars from hazardous roads to Western caves. Yet it makes surprisingly good time on the interstate en route to weekend caving trips. She deftly dodges traffic while talking almost non-stop on her cell phone, usually to other cavers about upcoming trips. In the three-hour trip from Boulder, Colorado, westward past Vail to Glenwood Springs early one morning in May 2000, she had just finished one call and was getting ready to dial another when the phone rang. The caller informed her that Hazel had been elected to the Board of Governors of the National Speleogicial Society. She would be filling one of four available seats, even as Nancy Aulenbach, whose term expired with the coming NSS convention, vacated hers. ■ "What?" Hazel screamed into the phone. "You mean I get to sit on the BOG? You know what that means in England, don't you? It's the same as sittin' on the loo." ■ It seemed a propitious beginning for the day's efforts, and I tried not to grimace as Hazel passed wide around an 18-wheeler at 85, howling about the BOG as we flew down the highway past the exit to Vail.

Hundreds of miles upstream of the Grand Canyon, a vast hot spring on the north bank of the Colorado constantly spews sulfur-laden, 124°

Above: Unusually white calcite formations line the walls of Fairy Cave in Glenwood Springs, Colorado. Right: Hazel admires a pristine formation chamber in one of the cave's recently opened tour sections.

water into the turbulent flow. In ancient times, some of this water collected in pools beside the river. These were barely cool enough for a person to enter. They became known among the Ute as possessing great curative powers. When the town of Glenwood Springs grew up around the pools in the 19th century, the Ute legend grew with it. Visitors arrived from ever increasing distances to soak in the mineral baths. The consumptive gunfighter Doc Holliday took comfort in the springs, as did outdoorsman and future president Teddy Roosevelt. Seeking to capitalize on the town's growing tourist trade, in 1886 a local attorney named Charles Darrow began offering occasional tours of a well-decorated cavern on Iron Mountain, 1,200 feet above the springs.

The bulk of the mountain was a lifted, tilted plateau of Leadville Limestone, a remnant of ancient seabed, unusual in that it contained high concentrations of iron, lead, silver, and gold, which was washed into the sea when the mountains vanished some 300 million years ago. The seafloor had gradually been pushed upward, and between seven and nine million years ago, the Colorado began cutting its way through the mountain, even as it began to carve the Grand Canyon. Fed by a deep geothermic source of heat, the original hot spring carved its way upward, out of the plateau and into the Colorado. As the last ice age came to a close 10,000 years ago, melting glaciers touched off a period of rapid erosion, with the river whittling away some 1,400 vertical feet of Iron Mountain. The hot springs soon found a new subterranean outlet to the lowered river. The older springs above dried out, leaving behind mineral deposits—and perhaps also stranding heat-loving microbes that became trapped within colorful secondary formations in the now dry cave.

Darrow named the dead springs the Fairy Caves, and built a path that wound through the 800 feet of passage then known to exist. The attraction offered little in the way of real competition to the saloons and gambling parlors crowding the downtown area. But these businesses catered mainly to miners and cowboys; in 1888, several entrepreneurs began working to bring a new class of clientele to Glenwood. A stone natatorium, bigger than a present-day Olympic-size pool, was opened at the springs, along with a well-apportioned bathhouse, casino, and "gentleman's club." Fifty feet away, work began on the Hotel Colorado, conceived as the largest and most opulent in the West. The rambling red stone structure, modeled after the 16th-century Italian castle Villa de Medici, opened in 1890 and was soon booking New York's elite for mountain vacations.

In 1895, with the hotel well established and the railroad on its way, Darrow formed the Fairy Caves Company and began remodeling his simple cavern tour. In 1897, he blasted a tunnel from the innermost chamber of the cave to the top of a 100-foot cliff in the mountainside. This new entrance gave a commanding view of the town, river, and canyons 1,200 feet below. The view would come as a pleasant surprise to patrons who entered the cave from the wooded road that wound up the mountain. He ran a wire down to the town's new hydroelectric plant, creating the world's first cave tour lit by electricity. By the time the railroad arrived later that year, Darrow could advertise the Fairy Caves—plural by dint of its multiple entrances and rooms—as "The Eighth Wonder of the World." The price of admission was 50 cents.

The Darrow family operated the tour until 1917, when the cave was closed to the public. Over the next few decades Darrow's improvements decayed. Only curious teenagers and the occasional caver entered the cave; many of the former broke off stalactites and other formations to carry home as souvenirs. But in 1952, members of the newly formed Colorado Grotto Club discovered new passages in the mountain that doubled the cave's known length and led to untouched formations larger and more beautiful than those at the heart of the

original tour. By 1960, members of the club had squeezed through a narrow slot at the back of the cave, named the Jam Crack, to find new levels, spacious rooms, and the profuse formations. Three cavers jointly purchased the property, intending to reopen the tour cave with routes to the new discoveries. They continued exploring and soon found yet another formation area, this one with stalactites and mineral draperies ranging in color from pure white to silver and black.

For the next two decades, Colorado cavers pushed back the known length of Fairy Caves, but the development plans never got off the ground. In 1982, a young caver and petroleum engineer named Steve Beckley became fascinated with the cave. He began trying to buy out the original purchasers, one

of whom, Peter Prebble, often camped in a shack outside the entrance. It took 16 years of negotiation, but in 1998 Beckley reached a lease agreement with Prebble. He began the difficult task of building an environmentally sound path to the new formation areas.

As with Arizona's Kartchner Caverns, area cavers pitched in to help Steve and his wife, Jeanne, create a tour highlighting the natural beauty of the cave. They took great pains to insure that virgin floors only a few feet away from the new trails were never touched by a single boot. They built elevated paths and stairways using boards made of recycled plastic, which would not decay in the cave or attract alien molds and fungi. They placed recessed lighting in short dark sections, that would be illuminated only when a tour

Hazel eases into a tight crawlway connecting two larger sections of Fairy Cave. The ancient thermal waters that carved the cave may continue to create new passages in active springs down the mountain, on the banks of the Colorado River.

group was present, thus keeping algae from growing on walls or formations. They created a second route for a wild caving tour that would let visitors safely crawl and squeeze their way into areas off of the main trail, seeing newfound rooms as the first explorers did.

One of the cavers helping with this work was Hazel Barton. A month after she returned from caving in Greenland, Hazel was with a team that discovered and mapped yet another heavily decorated chamber just off the wild cave route. They named it the Polar Bear. Not long after, a nearby room was found and named Discovery. The Beckleys decided to make the new room the highlight of the wild tour. A nearby chamber was crusted with spaghetti-like "pool fingers," the dried remnants of microbial growths in the ancients springs. Hazel named this room the Astro-biolab and made plans to return someday to sample the pool fingers for evidence of survivors—thermophile descendants that had somehow adapted to the drier, colder cave.

By the time the new tour opened to the public in the summer of 1999, Hazel had drafted a colorful map of the cave's three main levels for inclusion in the official guidebook, written by Jim Nelson, a local caver and historian. The guidebook

featured glossy color photos of the new formation. The cave's official new name was Glenwood Caverns and Historic Fairy Caves. The Beckleys had leased a gift shop and ticket office in town, where 2,200 visitors purchased tickets by the end of the first week.

The summer of 1999 was also when Hazel agreed to leave her teaching job in Denver in order to conduct postdoctorate research on tuberculosis at the Norm Pace Lab in Boulder. She had become a friend and admirer of Pace, and when he had moved his world-famous laboratory from Berkeley to Colorado University in Boulder, she jumped at the chance to work for him. Between the rigors of medical research and time spent mapping underwater caves in the Yucatán, Hazel had put off sampling the pool fingers of the Astro-biolab.

So it was that on a sunny Saturday in May 2000 I found myself stepping behind Hazel from the bright Colorado sunshine into the historic entrance of Glenwood Caverns. All of the cave's entrances were now gated with airlocks to protect the formations inside from the drying effects of the thin mountain air. As one of the two local cavers who had come along sealed the steel door behind us, Hazel checked over her sampling gear. Our plan was to remove several of the fossil pool fingers for genetic analysis in the Pace Lab, then go down the mountain to collect living pool fingers from below the waterline of the hot springs. Either location might reveal novel organisms. Neither had been previously examined by microbial researchers. But if bugs were found in both places, the interesting question would be whether any of them were closely related. If so, such a finding could indicate that heat-loving extremophiles were more resilient and adaptable than anyone had suspected.

I knew that Diana Northup, a microbiologist who had done a great deal of work in Lechuguilla, had also collected fingerlike

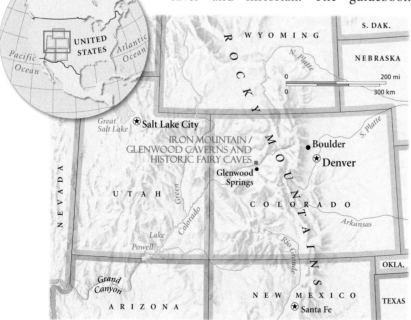

microbial mats from a sulfurous hot spring cave in Kentucky. She had analyzed the biological communities that made up these fingers at the Pace Lab, finding within them a sulfur-eating thermophile that, surprisingly, proved closely related to another known organism. Its closest living relative was a symbiotic bacterium that lived in the gills of the giant tubeworms lining the black smokers of the Pacific seafloor. Similar chemistry had helped the species establish toeholds in two vastly different extreme environments, located many thousands of miles apart from each other.

I wondered whether the Astro-biolab might hold the same sort of mysteries.

Our group of four walked down the tour route, following the Bright Angel Trail through a looping tunnel. The passage was named after a famous trail and creek in the Grand Canyon. On his first descent in 1869, Powell had called a particularly unpleasant spot in the Colorado the Dirty Devil. In order to offset the name, he decided on the second descent to dub a scenic creek, sporting a waterfall and a trail leading out of the Canyon to the South Rim, the Bright Angel.

The red and yellow hues of the cave passage were worthy of its namesake. I paused several times to admire the formations, which were among the most spectacular I had seen in a privately owned tour cave. Long soda straws appeared to drip milky and rose-colored rock onto terraced flowstone slopes of calcium carbonate so pure as to appear translucent. Red and orange bands of calcite called cave bacon rippled along the ceiling. Light sparkled from every surface, reflected by tiny crystal facets within the formations, never stained by mud from floods or cavers. After several hundred feet, we reached an intersection where the developed path turned right and a well-traveled passage turned left.

"This is where we get off," said Hazel. "We'll follow the wild tour route for a bit, so don't be surprised if we bump into some wild tourists."

The walking passage quickly became a crawl through boulders. After the mud of TAG, I was glad for the dry, relatively clean surfaces we scrambled over in the Colorado cave. We climbed up through a short slot and were soon in a walking passage again when I heard voices ahead—a group of the "wild" tourists Hazel had mentioned. We stopped and visited for a while. It turned out that all but one of the group of five were taking the wild tour for the second time, having tried it during the cave's opening season the year before.

"I had to come back to see if it was as fun as I remembered it," said one man, who this time had brought his teenage daughter along.

Shortly after we parted from the tourists, Hazel led me through a small opening, away from the wild cave route. I could tell that I was now moving through passage that had seen very little human traffic. Getting through required additional crawling and scrambling up short chimneys and down narrow slots. Several times Hazel or one of the others in our party turned to let me know that we were passing under a formation area, so I should keep my helmet low to avoid damaging soda straws or thin curtains of banded calcite.

Within 20 minutes we emerged into an irregular chamber about the size of a suburban living room: the Astro-biolab. Parts of the ceiling were lined with globular formations known as "cave clouds." I had seen similar formations in pools in Lechuguilla, and it was easy to believe the two caves had been formed by the similar action of rising hot springs. Delicate, lacey aragonite crystals filled many cracks and crevices in the wall; others grew outward precariously from soda straws and stalactites. On the ceiling were tiny white dots that glistened with condensed moisture. I had learned from Larry Mallory that such dots were often the visible growth a bacterial colony.

"Well, here we are," said Hazel, distracting me from my inspection of the wall.

I followed up a large boulder into a higher

part of the room. She pointed out a refrigerator-size swath of wall that seemed to be made of rusted vermicelli. These were fossilized pool fingers, clumped together in a stone mat in a way that looked as though the water had drained from the room just yesterday. I had seen living fingers like these in the stream that passed through Cueva de Villa Luz, a Mexican cave full of poisonous hydrogen sulfide gas that nonetheless sported a thriving sulfur-based ecology. There, other stringlike microbial masses called "snottites" hung from the walls, dripping sulfuric acid that rapidly eroded the limestone to make new cave passage.

Hazel pulled out her kit and we both donned latex gloves. As I had done with other microbiologists, I assisted by flaming the needle-nose pliers she would use to pry out bits of the petrified spaghetti: I held out a bottle of ethanol, and she dipped the tool into it. Before the alcohol could evaporate,

I ignited the pliers with a cigarette lighter. Hazel quickly removed a sample from the wall with the sterilized tool. The thin rock crunched like a pastry crust as she pulled a strand loose. She dropped the stony bits into a vial containing a solution designed to preserve the DNA of any living organisms. We recorded the sample location and number on the bottle with a permanent marker, and then repeated the process. Within a short time, we had filled all six sample bottles that Hazel had brought into the cave. Mission accomplished, we packed and left.

As I grunted and groaned through crawlways, several aches and pains leftover from the Hog Wallow renewed themselves. Trying to keep up with Hazel and two other twentysomething cavers reminded me of my approaching 41st birthday. I was on the verge of asking them to stop and rest a bit when we suddenly popped out on the tourist trail. Once back on the main walkway we detoured to

Hazel exits from an artificial entrance to Glenwood Caverns, created to allow tourists access to newfound formation areas. The cave's tour routes offer easy access to wild areas with the potential for unexplored passages.

enjoy the view from Exclamation Point—Charles Darrow's excavated overlook—and to see a marvelous formation area called King's Row in a spacious chamber dubbed the Barn. It was still early afternoon when we stepped outside, and we decided to drive down the mountain to sample pool fingers from the hot springs before having a late lunch.

The water at Glenwood Springs emerges from several natural openings into a deep blue pool hot enough to kill a person in minutes, should anyone be unfortunate enough to fall in. The pool is fenced and gated, located just east of the large bathhouse and bathing pool, both of which still enjoy booming trade. Water from the pool flows through a cooling system and two decorative fountains before entering the bathing area, which itself is graded into pools of warmer and cooler water to suit the tastes of various bathers. The springs management had granted Hazel permission to collect samples, and the operations manager accompanied us into the natural spring enclosure.

Along the edges of the pool, white and green tendrils waved in the current. The green fingers resembled a stringy plant; the white ones a pale root. But I knew from reading NASA research on the hot springs of Yellowstone that neither were plants at all, but colonies of many species of microorganisms. The green fingers were likely to contain photosynthesizing bacteria, as opposed to true plants. When plants first evolved from eukaryotic microbes 500 million years ago, they did so by absorbing a number of these sunlight-eating bacteria—which had existed on Earth for over three billion years—into their much larger cells. The once free-living bacteria became chloroplasts, the energy producing structures that color all plants green. Yet research at the Pace Lab and elsewhere had shown that chloroplasts still retained some of their original genetic code, separate from the DNA of the plant cells into which they are incorporated. This genetic material proved that the chloroplasts of all plants on Earth, from redwood trees to spinach, are close

A RUDIMENT FOR CAVERS

BY TIM CAHILL

"Check me out here…" I'd typically ask my caving buddies before I rappelled into some abysmal darkness half a mile deep into the Earth, "am I death rigged?"

To rappel three or four hundred feet down a rope, which is anchored somewhere behind you, you hook onto that rope with a device known as a Thor rack, an attenuated horseshoe-looking gadget with numerous gates. It hangs from the top of your seat harness, looking oddly obscene. The climbing rope is run through the gates in the rack in the way that a skier slips around slalom poles. If you thread the rope backward, when you lean out over the lip of the precipice, all the gates snap open and you go hurtling to your death wondering why you didn't ask someone to check out your rig—even though you've threaded the rack a thousand times before.

Cavers don't mind checking to see if you're death rigged. Better to check the rack than pull a dead body out of some great cave, thinking every time you go in again about who died there, and why.

There are many ways to die in a cave: you can drown, freeze, fall, be crushed by falling rock, or simply get lost and starve to death. This list does not exhaust peril's possibilities. Cavers do not talk about the rigors or splendors of their sport to outsiders, in part because, typically, they are the ones who have to affect the rescues or, worse, retrieve the bodies.

As a caver, you'll come into some motel after 18 hours of crawling through the bowels of the earth, still wearing your muddy coveralls and filthy boots. Someone will ask what you've been doing, and if you don't actually lie—"We've been cleaning out sewers"—you may actually say that

Chris Stine swims past packed formations into a new discovery in Lechuguilla.

you've been caving.

"Oh," the person says, trying to establish rapport, "spelunking."

Cavers haven't called themselves spelunkers for almost 40 years now. They may belong to the National Speleological Society, but they don't spelunk. They cave. The sport is caving.

"So where do you guys go?"

The answers you give are vague, or misleading, or downright lies. As a caver, you're not going to tell novices where you've been and how to get there, because those persons just might attempt an exploration. They'll take a flashlight and a clothesline and one or the other will break or malfunction and then they'll die or have to be rescued,

and you'll be responsible. Alternately, they may leave cigarette butts behind or candy wrappers; take a souvenir, some small geological wonder that took five centuries to form, foul a pristine pool with soap, or leave feces that real cavers pack out; they may go tromping over some delicate flowstone that requires a barefoot traverse or spray-paint their name on the wall.

These are some of the reasons why cavers are secretive, discourteous, and sometimes downright rude to folks who ask casual questions about caves. They are not going to tell anyone about the places they hold sacred. It's rather like a religion and caves are the cathedrals.

So what do you do if you are a novice and are sincerely interested in caves? Your best bet is to contact the National Speleological Society, 2813 Cave Street, Huntsville, Alabama, 35810 (www.caves.org). They will get you in touch with your local chapter, called a grotto. You'll go out on guided expeditions, learn a lot about caving and cave conservation, and you'll begin to understand the dangers. In time, you may become one of the 10,000 or so hardcore cavers in the U.S. Then when someone asks you how to find one of your favorite caves because he wants to go "spelunking," you'll likely tell a lie and send the poor fellow off to some vast and caveless field in the middle of nowhere. You're a caver now, and you have your rude responsibilities.

cousins to photosynthesizing bugs that still thrive in hot springs. What evolutionary theorists of previous generations had considered the "rise of life on Earth," beginning with the spread of plants, was only a very late chapter in a microbial story that had unfolded far earlier. The vast history of life on the planet, and the most powerful and persistent life of the planet even today, was overwhelmingly microbial.

Dropping our hands quickly in and out of the water, taking extreme care not to fall in, Hazel and I each collected several fingers from the edge of the main pool. A few feet away, on slightly higher ground just inside the surrounding fence, Hazel noticed a two-foot wide opening above a small, steaming grotto. She pulled a flashlight from her pack and shined it inside: a small spring flowed along the bottom toward the main pool. She dropped down into the five-foot-diameter steam room and discovered some white pool fingers at its edge. I stood at the entrance and passed two sample bottles down to her so

that she could collect of these; she quickly passed the full bottles back and asked me to label them Dark Zone 1 and Dark Zone 2.

"You look more British than ever with those pink cheeks," I joked as she climbed out of the steaming hole, sweating and nearly out of breath after only five minutes in the extreme heat.

Hazel collected additional pool fingers from a decorative fountain, and then the four of us rewarded ourselves with a luxurious meal at the Hotel Colorado, followed by a soak in the hot springs pool. All in all, it was far more civilized manner of ending a day's caving than the TAG practice of hiking down the mountainside to find the nearest greasy spoon.

Over the next two days, Hazel gave me a crash course in the 16S subunit of RNA and the ways this molecule could be coaxed from a biological sample to show it's evolutionary relationship of all living things on Earth. Extraction was a complex process,

Left: The large pool at Glenwood Springs draws tourists to soak in the sulfurous thermal waters that emerge on the banks of the Colorado. Above: Hazel samples "pool fingers"— strands of thermophilic bacteria— from one of the springs that feed the pool.

Above: Hazel collects
fossilized "pool fin-
gers" from a wall of
Fairy Cave. The hot
springs that deposited
these fossils receded
thousands of years
ago, but perhaps
some of the life
adapted to the drier
cave and still sur-
vives. Below: Famed
microbiologist Norm
Pace discusses the
protocols Hazel
will use in genetic
investigation of her
Glenwood samples.

involving dozens of steps in which the genetic material could be purified, amplified, and analyzed. Each of these steps had been perfected on a large scale by researchers in Norm Pace's lab through years of trial and error—the various recipes to be used on different types of samples were kept in a large loose-leaf notebook that formed the heart of the Pace Lab.

As complicated as it would be to get any information from the hot spring samples in two days, Hazel explained, it would be a far more difficult task for us to extract useful genetic material from the cave samples. The heavy calcite coating that made the cave samples solid could also seal away the DNA of any organisms within the fossilized fingers. It would take a variety of solvents and other treatments to "wash" genetic material out from the rock in useful quantities. So for now, Hazel concentrated on the wiggly bits of wormlike fingers from the spring pools.

As Hazel extracted the first sample beneath a hooded workbench, she mentioned idly, "This is a lot easier than getting TB cells from diseased lung tissue, which is what I'm usually doing in here."

She would perform each step in the preparation of a sample on one of the main pool fingers and another from the display spring. I would then copy the procedure with one of the Dark Zone samples and another from the main spring. Ideally, we would be able after two days to get a rough idea of the number of organisms the sample held from each of the three main branches of life: the archea, the bacteria, and the eukaria. We would also be able to tell if we had contaminated the samples in the lab or had otherwise erred in any of the many steps of preparation. Finding the type and species of individual organisms within any of the samples would take days or weeks longer, involving still more complex lab procedures.

The final step of the initial amplification was to place a drop of prepared material into holes in a gelatinous brick. Over a period of several hours an electric current would pull a long dotted line from the holes across the gelatinous surface. The bars and dashes of this line would indicate the type of life within, as well as any evidence (if we had messed up) of contamination during sample preparation. I felt like a proud father when, on the evening of the second day, my Dark Zone sample had shown clear, definite bands in all three branches; the blank controls showed no sign on contamination. Even if Hazel had done the hard part, climbing into the little steaming cave to extract its fingers, this was "my" sample, lovingly ushered along through long hours in the lab.

By connecting the evolutionary history of hundreds of types of organisms from all three branches, Norm Pace and his

colleagues have filled in much of what he calls "The Big Tree"—the diagram showing the genetic relationship of all life on the planet. The tip of one twig near the end of the eukaryotic branch held all multicellular life (plants, insects, toadstools, fish, animals, us), but the vast majority of all three branches display the diversity of the microbial world. While I was in the lab, Norm gave me a Pace Lab T-shirt, prized among scientists, which shows the Big Tree, with an arrow pointing to the tip of the eukayortic branch indicating "You are here."

Genetically speaking, there is little difference between a human and cabbage compared to the far greater genetic distance between, say, *E. coli* and the bacteria that have been identified in caves. Occasionally, microbial samples from extreme or obscure environments lead to new twigs on the Big Tree at the kingdom level—the level at which plants and animals differ from one another. Pace explained to me that he thinks field microbiologists who attempt to retrieve live cultures from extreme environments

without first doing a shotgun genetic study are putting the cart before the horse. They may retrieve one or two or even a dozen new microbes, but it is the hundreds or thousands of organisms in a particular environment that, taken together, can explain the operative chemistry and diversity of the entire ecosystem. A single bug offers a very incomplete—and potentially misleading—glimpse of that larger picture.

Hazel was excited to keep working on our samples, to see what organisms lay within the Dark Zone and to begin the slow extraction process on the cave samples.

"It's good stuff in there, I'm sure of it," she told me. "But," she warned, "I may not be able to work on it until after the Grand Canyon expedition and the NSS convention in West Virginia. And I've got a lot of grant work with my TB samples to do in the lab. So it may be this fall sometime before we find what we've got."

I nodded. At least we had intact samples in the lab. The info was there, waiting to be extracted.

Hazel begins the slow process by which the genetic fingerprints of new-found microorganisms are amplified and ultimately identified. She and others in the Pace Lab can spends weeks separating and typing genetic lines from a single field sample.

CHAPTER 12

THE HIDDEN REALM

Rappelling in 112° heat with the relentless Arizona sun bouncing from every rock in the 700 feet of exposed space below her was both exciting and breathtaking. Nancy was not looking forward to having to do a 300-plus foot climb back up this cliff face after she and Hazel explored the cave. Like most cavers, she prided herself on her abilities to ascend under her own power from any place she was able to reach. But that ride out would be a long way off. As she slid downward

Above: Dark shades and sunscreen protect Hazel from the sun while she awaits her turn on rope. Right: Nancy enjoys the view of the calcite-laden river far below as she descends to the IMAX Cave.

along the wall, the entrance that had appeared little more than a shadow from the Little Colorado River below now grew to impressive dimensions, perhaps 20 feet wide and 10 feet tall. But for its inaccessibility, such an obvious entrance would have been mapped many years earlier. ■ The cave was recessed from the cliff. To reach it, Nancy had to swing back and forth like Tarzan until she was able to grab a boulder and pull herself in. Before completely derigging her gear from the rope, she attached a smaller line to it so that she could pull Hazel in when she rappelled next and, more importantly, so that both could reach the line out when they were done exploring. The task accomplished, Nancy looked up in the general direction of the ledge, now obscured by the bulging cliff face, and hollered, "Off rope!"—Hazel's signal that it was clear for her to come on down.

Filmmakers rigged two lines down the cliff face to the oval-shaped entrance in order to accommodate the large crews that would be making the 350-foot trip straight down to the cave and up again throughout two days of filming. Nancy, standing, began to map the entrance area almost as soon as she stepped off rope.

While she waited, Nancy walked a short distance into the darkness, letting her eyes adjust from the incredibly bright Arizona sun to the round beam of her cap lamp. As large as the entrance was, the first chamber was larger still. After clambering over an enormous breakdown block, she could see that the passage belled out to a width of about six feet, with the ceiling reaching 30 feet or more overhead. The large breakdown chamber stretched over 100 feet in the distance. The walls sparkled with dogtooth spar, calcite crystals with one-inch triangular faces, thick enough in places to rival those of Jewel Cave in South Dakota. Their presence suggested that the cave had been flooded for long periods of time in the distant past. There appeared to be some small tunnels leading out of this chamber along the right-hand wall, but before she could take a closer look, she heard a voice behind her.

"Hello in there," Hazel called.

Nancy returned to the entrance and quickly pulled Hazel in to join her. They marked the rope so that they could later measure the distance from the ledge to the cave—283 feet—then chose a centrally located boulder in the entrance from which to begin their survey. The first survey shots in such a large chamber, using tape, compass, and clinometer, should go pretty quickly, in that they could move forward 50 feet between stations each time. And the process would have been quick, if they hadn't been asked to backtrack and repeat their rappel several times for Steve Judson, Brad Ohlund, Jack Tankard, Barry Oswick, and other seasoned members of the film crew who soon joined them below.

Until now, Steve and Brad had agreed that Greenland was by far the most grueling location they had ever visited, but Greenland had never offered a moment of intense high anxiety such as they experienced above the canyon of the Little Colorado. Even while filming on Everest, they had never felt so exposed to a sheer vertical drop as they were here, where a peregrine falcon zipped past Steve as he descended. The bird spiraled outward and below him toward the river 2,000 feet below. On one of the later rappels, aerial photographer Ron Goodman captured Nancy from a helicopter as a tiny, spiderlike figure before an immense cliff in what would become the centerpiece of film's opening sequence. Later that day, Steve was able to helicopter the Navajo family that had done the sand painting through the same path, to enjoy a perspective of their ancestral canyon that they had never seen. For its ability to inspire vertigo, this rappel beat far deeper—but darker—drops in Mexico and TAG.

It wasn't until they helicoptered back to the ledge the next day that Nancy and Hazel were able to begin mapping in earnest, accompanied once again by the film crew. After measuring the entrance, they discovered that the bottle-shaped main chamber was six feet wide by 135 feet long. Hazel dropped into one of the small holes along the right-hand wall, a triangle that led to a crawlway about 30 feet long. At the far end of the crawlway, she could see blackness indicating that it opened up.

"Bad news," Nancy called down to her. "Barry just hollered that they need us back at the entrance. You need to get back on rope."

Hazel turned to look back out of the triangle. "Pretend you can't hear him. I think this goes."

She then performed one of the first acts of any caver in a new passage: holding up a hand as though greeting the cave ahead. But rather than the touching ritual it seemed to be, this gesture was a way to determine the size of the cave. In smaller passages like this one, even a faint breeze would indicate that a massive volume of air was moving from somewhere beyond, equalizing the air pressure differential between unknown passages beyond and the surface. Canyon caves are often little more than rock shelters—recessed overhangs that end within a hundred feet or less of their entrances. But here a definite whisper of cool air pushed against her outstretched palm.

"We've got air," she said to Nancy.

"Ye hah!" Nancy said, as she headed into the passage with Hazel."

Hazel slithered down the crawlway. The dimensions definitely got larger 20 feet or so ahead, and the breeze in her face felt stronger.

This was the sort of airflow she had felt before in cave systems that turned out to be many miles in length.

"Come on girls, you're needed now." This time the voice behind them was Barry's, shouting from the entrance. They took a last longing glance at the going lead, and then headed out for more filming.

Water and sunlight feed a plant, but in the sunless world of subsurface microbes, the key ingredients for life are water, carbon, and minerals. Thus to find a rich area for new cave bugs, one has to look closely for places where water and minerals meet.

I once saw Larry Mallory give a talk on

Hazel sketches room details for the first map of the new cave, which, despite its nerve-wracking entrance, proves to contain only horizontal passage. Because the main chamber presents relatively easy walking passage, she wears her vertical gear while mapping in order to avoid having to stow it and then don it again when climbing back to the top of the cliff.

Cave-dwelling Bats:
Misunderstood but Vital Allies

BY MERLIN D. TUTTLE, PH.D.

Contrary to popular misconceptions, bats are gentle animals and are both beneficial and necessary to the balance of nature worldwide. Bats pollinate the planet's peaches and bananas; without them there would be no tequila or kapok, both of which are made from bat-pollinated plants. Their guano is among the most productive natural fertilizers known and highly prized by farmers and gardeners, and they are the primary predators of night-flying insects, which includes pests that if not for the hungering bats would cost farmers and foresters billions of dollars annually. In tropical regions, fruit and nectar-eating bats disperse seeds and pollinate flowers vital to the survival of rain forests and to the production of crops worth hundreds of millions of dollars each year.

While most people associate bats with caves, nearly 1,000 species worldwide can be found living in almost every habitat, from deserts to rainforests. Bats do use caves for raising their young and hibernating in winter. Spectacularly large colonies of bats live in a few well-known caves, but the vast majority of caves are not suitable for bat use. Only one in hundreds, or even thousands, of seemingly available caves will actually meets bats' needs, making the appropriate caves that do exist critically important to their survival. Bracken Cave in Central Texas, for example, shelters the world's largest bat colony—some 20 million

Mexican free-tailed bats (*Tadarida brasiliensis*)—located just north of San Antonio, the cave is a vital nursery site for the species. Nevertheless, even these bats must migrate up to 1,000

Odd facial features of pocketed free-tailed bats (above) assist in radar; the huge ears of the California leaf-nosed bat (below) picks up returning echolation signals.

miles or more to the south to find other caves equally fit for wintering in Mexico. Bats are extremely attached to their caves of birth or hibernation and have few other options suitable for their highly specialized needs.

In this sense the survival of bats seems fragile, dependent as they are on scarce natural resources. Unfortunately their vulnerability extends well beyond habitat suitability and supply. Bats are among the slowest reproducing mammals on Earth—most bats produce only one pup per year. Disturbances to their cave habitats are known to be the primary cause of population declines. Also, congregating in large numbers, as they do in caves, when endangered, they can be exposed to large-scale destruction, extinction, rather than the relatively less devastating individual, or small population, reductions.

Clearly cavers can endanger bats. A single caver passing through a hibernation cave can cause thousands of bats to awaken prematurely, costing them 60 days or more of stored fat reserve— a supply that must last until spring if the bat is to avoid starvation. Even when undisturbed, hibernating bats will occasionally wake up during the winter and move about; they must drink water because their thin wings, ears, and tails constantly lose moisture to the cave air. By the time spring arrives, an emaciated bat may have lost up to a third of its body weight, and must feed immediately. With too many winter interruptions, bats will not make it to spring at all.

Like humans and other creatures producing few offspring, bats are excellent caregivers for their young. When an infant bat is born, it clings

Some 20 million Mexican free-tailed bats—the world's largest bat colony—exit Bracken Cave in central Texas each summer night to feed on mosquitoes and other insects throughout the region.

to its mother for up to a week, feeding as she flies and roosts. When it's strong enough to cling to a cave ceiling on its own, the mother bat will exit the cave to feed alone, leaving the infant in care of other "baby-sitting" females who watch over the nursery. Upon her return, the mother can find her own infant among thousands or even millions of others in the cave. She recognizes both the smell and the cry of her offspring. However, if humans upset a nursery colony while the mother is away, the disruption may keep her from locating her baby.

It is easy to understand, then, why bat conservation places protection of their critical roosting caves high on the list of priorities. It is equally important to teach people about bats and their needs. On the simplest level, we all need to know that bats are not blind, bats do not become entangled in hair, and even bats that are sick rarely bite unless handled. Left alone, bats are harmless and essential members of their ecological communities

The easiest approach for anyone to take is to encourage and protect bats in your own neighborhoods. Many people study and benefit from their local bats by building specially designed bat houses in their yards and gardens. Once bats take up residence in a bat house, they will consume insects on summer nights and create guano that you can use to increase the yields of your spring gardens. In winter, your bats may depart for some distant hibernation cave, but they will often return year after year. Both ready-made kits and plans for bat houses are available from many nature stores, museums, and conservation organizations.

Bat Conservation International, which I founded in 1982, is a nonprofit organization dedicated to bat conservation and the ecosystems that rely on them worldwide. The organization now has over 14,000 members in 75 countries, and thanks to widespread support, private corporations, government agencies, other wildlife organizations, and many cave explorers are now collaborating to help bats. Bat Conservation International's website *www.batcon.org* includes a wide variety of information on bats and their conservation needs, extensive bibliographies and reading lists to further your knowledge, and practical tips on building bat houses. Information can also be obtained by calling 512-327-9721, or writing to Bat Conservation International, P.O. Box 162603, Austin, Texas 78716.

hunting cave bugs to a group of rocket scientists at NASA's Jet Propulsion Lab who were attempting to design life-detecting probes for future missions to the Jovian moon Europa. In the talk, he displayed a close-up slide of helictite from Lechuguilla cave, a curlicue of calcite with a silvery drop of water clinging to its tip. "When I collected here"—Larry used a pointer to indicate a spot on the helictite no more than a centimeter from the tip—"I was able to recover one organism. When I collected here"—he pointed to the water droplet—"I was able to collect over 400 separate taxa, with many more individual species of microbes. To a microbe, a centimeter is a huge distance. If you want the best results, you have to look for the places where food, in the form of minerals, meets water, which lets the bugs do chemistry with their food."

This chamber, which Hazel had christened IMAX Cave in her survey book, was extraordinarily dry and dusty. Yet it held ample evidence of water in its past, so she began searching the walls for places where vanishing water might have clung for the longest period of time. It seemed unlikely that any microbes would be actively growing in such dry conditions, but "actively growing" and "dead" are very different things to a bacterium. Microbes in other extremely dry environments—Death Valley, the dry valleys of Antarctica, and buried salt deposits—had been shown to enter a state resembling suspended animation as they became dehydrated. Like the "instant Martians" of the Bugs Bunny cartoon, a drop of water could bring them back to life, even many years later. Studies of primitive bacteria preserved in deep underground salt deposits in New

Mexico suggested that such organisms could be revived after an incredible 250 million years of inaction.

Although Hazel doubted much microbial metabolism was taking place, she didn't doubt at all that DNA of new organisms might be lying on one of the walls in front of her. She decided to collect some material from downward-projecting tips of calcite spar; during moister times, each of these would have had at least a drop of moisture clinging to it. With Nancy assisting, she also collected from places in the rock wall where it appeared two different minerals came together—any juncture of differing rock chemistry could be a rich stamping ground for life.

Tullis Onstott, a geomicrobiologist from Princeton, had argued that in some cases veins of differing minerals within a rock face not only attracted life, but also were the direct result of it, providing visual evidence of microbial communities altering their surroundings. He had descended two miles underground into the world's deepest gold mine in South Africa, a hellish environment where even with cool air blown in from the surface the temperature exceeds 140° F. There he had found that the greatest concentrations of native microbes were directly tied to the greatest concentrations of gold and other metals. Onstott theorized that the veins of pure gold in the mine were actually deposited by the bugs as they gathered the metal from the surrounding rock matrix through millennia of chemical metabolism. As a result of such discoveries by Onstott and others, microbes cultured from deep mines were now being used with great success to collect "lost" gold from old mine tailings. Not only were extremophiles natural chemists, they were natural miners as well.

Hazel carefully labeled and preserved each sample she gathered for the camera in IMAX Cave. She suspected it might be months or years before she actually processed these, as she was more immediately interested in examining the samples from Glenwood caverns, in collecting and examining new sam-

ples from the halocline in the Yucatán, and in continuing her medical tuberculosis research. But she knew that properly labeled and stored, these samples would wait for her, whether she decided to look at them next month or in 2045. Just as the blowing, going lead in the chamber beyond would keep—the crew had to meet the helicopter above in an hour, and they would not be coming to IMAX Cave on this short expedition—she knew that whatever dormant life clung now to her cotton swab would remain until she was ready to delve into its secrets.

The final sample stored, the final shot taken, the crew hastened to remove the ropes and cranes, pulling all evidence from cave

Nancy prepares to empty a long sampling tube into a storage container. By analyzing the percentage of suspended calcium carbonate in the water of the Little Colorado, geologists might be able to predict the potential size of the unexplored caves from which springs in the river issue.

and cliff that they had been there at all, as they had promised the Navajo they would do.

The summer passed. On the Little Colorado River, Hazel had spotted a boiling spring, the underwater cave that was the source of the milky calcite that colored the water turquoise. After the Grand Canyon trip ended, she learned that divers had entered the spring just once before—a very short two-man scout led by the late Sheck Exley. He had reported a large going passage, one that would require a full expedition, with many tanks and a support crew, to explore.

At the NSS convention in June, Hazel and Nancy began developing plans with Donald Davis and Yucatán cave diver Dan Lins to return to the Little Colorado in the spring of 2001. From a base camp on the banks, a spot where Powell and his team had parked their dories and passed a pleasant evening during the second Grand Canyon expedition of 1871, separate crews could explore both the spring and IMAX Cave. I immediately asked if I could come along.

Several weeks later, Hazel called to give me more details of the developing expedition. Almost as an aside, she added, "Of course, between the dive trip I'm about to do in the Yucatán and all the exciting stuff with the Glenwood samples, it will be a while before we can firm up the Grand Canyon dates."

"What exciting stuff with Glenwood samples?" I asked.

"Didn't I tell you? Your Dark Zone sample has 37 new species of bacteria—two of them at the kingdom level. It's amazing, like finding two new classes of life, each as big and different as 'plants' or 'fungi' are from everything else. Norm's terribly excited. And we haven't even looked at the archea in the sample, let alone touched any of those we got out of the cave. It's very big stuff."

It was indeed. Yet as I hung up the phone, I couldn't help but hope, with an explorer's eternal optimism, that the biggest, most amazing stuff was yet to come.

Leaving at least one unexplored passage behind for now, Nancy takes a parting look at the IMAX Cave on the day that humans first set foot inside it. Hazel plans to lead a return expedition to the cave in the near future.

203

During the filming of *Everest* in 1996, I slept on the ice floe at the base of Mount Everest, which taught me one thing: the most valuable asset on any glacier is that inch and a half of foam separating your bones from the slab of hard ice. When I went to Greenland to film *Journey Into Amazing Caves*, I had the good sense to bring along an extra sleeping pad, so I was snoring like a cave man when a storm slammed into our camp around midnight.

Gusts of 60 knots, then 80 and higher lashed my tent. It was like being trapped inside a snare drum during a rock concert. I knew that the nylon would be ripped to shreds in moments so I stuffed my boots and other essentials into my sleeping bag. When my tent explodes like a blown-out spinnaker, I thought, at least I won't be left to face the night in just my long johns.

The tent gripped the ice for dear life. I pulled my sleeping bag over my head to muffle the deafening roar outside and finally drifted off to sleep, thinking, I could have been an accountant. I could be in a warm house, under an electric blanket.

It's a crazy way to make a living, wrestling bulky camera gear into places it doesn't want to be. Here we were, a team of 15 exhausted people about to focus every ounce of energy onto the crucial yet absurd task of capturing on film the way light glinted off a cave wall. What for? So that strangers in a darkened theater would experience an illusion: they would feel that they had been somewhere they might otherwise never have a chance to go. Putting yourself through a hardship like the Greenland shoot makes no sense, unless you are obsessed with bringing to people astounding images that will touch and exhilarate them—and then it makes all the sense in the world.

This way of life requires passion.

And so it is with cavers. When I began the research for *Journey into Amazing Caves*, all the cavers I talked to radiated passion about their encounters with the underworld and the urgent need to protect it. Here was a group of zany zealots in the grip of an obsession their friends and families often couldn't understand. I wondered if we could capture that intensity on film. I wondered if we could convey what it is about caves that inspires such passion, a feeling I experienced first-hand in Greenland when Janot Lamberton led me on that initial descent into an ice cave so blue it took my breath away.

To reveal the full beauty of these magnificent caves, I knew we would have to bring together an exceptional team of filmmakers. Filmmaking, like caving, is a team effort. I knew the film would stand or fall based on the chemistry—or lack of it—between all of the people involved. Some of us had already worked together. This was the 20th large-format film Producer Greg MacGillivray and I had shaped together. Cameraman Brad Ohlund and I had collaborated since 1991, when we stood side by side in the path of an oncoming tornado, for the questionable purpose of filming it with an IMAX camera. Others, like Nancy and Hazel, barely knew each other. Wes Skiles, the best underwater cave photographer in the world, had never worked with Howard Hall, the best large-format underwater cameraman—and now they would have to risk their lives together to bring back images of the underwater caves that lay far beneath the jungles of the Yucatán. Somehow, it all clicked. Maybe it was dumb luck. But I came to believe it was the power and focus of shared passions.

This film presented enormous technical and logistical obstacles, but for me, personally, each shoot had its defining moment. In Greenland, that moment came when the Brown brothers teetered on an unexpectedly precarious block of ice at the bottom of a 500 foot cave, and radioed back their decision—they would stand their ground, risks and all, and keep filming. In the Grand Canyon, the moment came when I looked down while on rope at the milky ribbon of river hanging 1,200 feet below my dangling feet, and saw beneath me a bird of prey rising in lazy circles toward the cave we were filming. In the Yucatán, the turning point was the comical flash of inspiration to use trash can lids to control the direction of the underwater lighting, a humble ploy that created a cinematic illusion of the utmost sophistication.

Each night in Mexico we all crowded into Wes's hotel room to watch video of the underwater footage he and Howard had captured that day. During these sessions I often felt swept along through the caves by an accelerating current, which was nothing more than my own growing conviction that *Journey Into Amazing Caves* could give people something they rarely get—a glimpse into an utterly unfamiliar dimension of their own planet.

The spectacular images we were capturing on film would be even more powerful, I suspected, if we could also tap into the long-buried emotional connection we all have with caves, whether we know it or not. During one of many marathon phone calls, the film's ever patient screenwriter, Jack Stephens, listened to a raw, gristled bone of an idea from me. If our ancient ancestors were cave dwellers, I speculated, then are caves not highly charged places connected to our collective subconscious? Most ancient cultures understood this. The Maya believed that many who ventured into the underworld perished, and that those initiates who did return came back transformed, enhanced by the power and beauty of the underworld and blessed with insights to share.

So it is, still.

Cavers are today's courageous messengers from the underworld, transformed by its spell, initiates of an uncharted world full of secrets, where discoveries that improve the human condition may be found. For some, the discoveries are as simple as glimpsing the beauty of nature in a roomful of glistening cave formations previously hidden throughout all of human history. For others, the discoveries become as difficult and complex as retrieving bizarre microbes from remote habitats, then using them to create new cures for human disease. Perhaps the magic of modern science is not so far removed from the magic of ancient peoples after all.

Jack and I agreed that my idea was too speculative to state outright, but still it formed the emotional heart of the film. It became both my job and my passion to share this strange new world with our audiences, as they sit in darkened IMAX theaters so evocative of caves: to make them feel the primordial spell of the underworld and its power to transform us all.

STEVE JUDSON

Steve Judson is the Director of Journey Into Amazing Caves, *the 25th large-format film produced by MacGillivray Freeman Films. Since 1988, Judson has edited almost all the large-format films from MacGillivray Freeman Films. He has also co-directed and co-written several films, including* Everest, *a film that took him on a journey in the opposite direction from the subterreanen realms he explored in* Journey Into Amazing Caves. *He was, he likes to joke, especially ill-suited to work on a film about caves as he suffers from claustrophobia.*

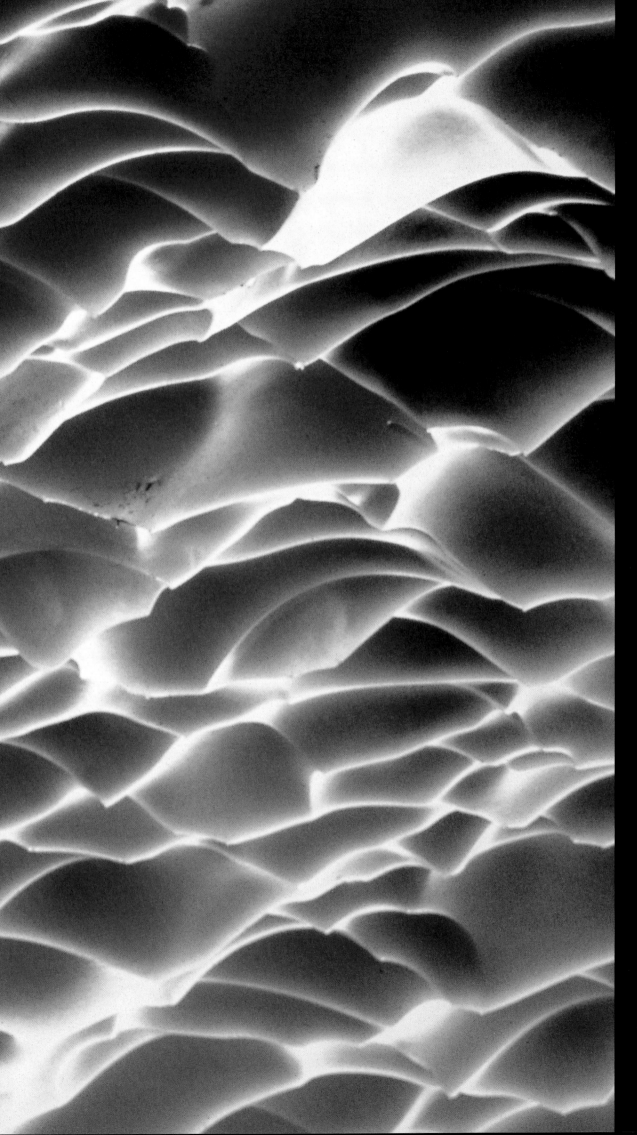

Specially designed ice screws anchor German caver Stefan Geissler 50 feet above the floor of a glacial cave in southern Germany. Most limestone and lava tube cavers try to avoid dangerous sport climbs—and the requisite anchors that can permanently mar pristine walls. But the short-lived nature of an ice cave makes it a perfect playground in which to hone climbing skills.

ADVICE FOR
FIRST-TIME CAVERS

BY MICHAEL RAY TAYLOR

The only safe way to try caving is to go with someone experienced.

Luckily, there are many ways for the beginner to do this. Most national and state parks featuring caves offer guided "wild cave" tours and adventures, as do many privately owned tour caverns. Such commercial trips commonly provide beginners with a caving helmet and lamp, limited group size, pretrip orientation, and trained leaders to take your group through the cave. Wild tours at some of the better known caves—such as Mammoth and Carlsbad—are especially popular and may require reservations weeks or months in advance.

Another safe way to enter a wild cave is to sign up for a novice trip with a local cave club. Clubs affiliated with the National Speleological Society, called "grottos," often schedule beginner trips and training events. Some will also arrange special trips for scout groups, schools, and similar organizations. For those who may have tried commercially guided tours and found them to their liking, joining a grotto can provide advanced training and the opportunity to take part in more difficult trips. Many grottos offer specialized workshops for members in mapmaking, cave conservation, and vertical techniques. Grottos are often designated caretakers of gated caves within their region and maintain keys, directions, and proper sign-up procedures necessary to visit gated caves. To find the grotto nearest you, visit the NSS online at www.caves.org or call the NSS national office in Huntsville, Alabama, at (256) 852-1300.

Grottos will usually provide helmet, lamps, kneepads, and other gear for beginners, although they may require you to purchase batteries or provide backup lights, water bottles, etc. Whether you go with a cave club or a commercial tour, for your first trip underground you should wear sturdy clothing with long sleeves and boots with lug soles and good ankle protections (running shoes and the like are virtually useless on slick, muddy surfaces). You should carry one or two small pocket flashlights as backups to whatever lights are provided—three separate light sources per person is the minimum for safety—and you should carry at least one change of batteries for each light. Gardening or work gloves will protect your hands from rough surfaces while crawling and climbing, and a personal water bottle, carried in a nylon day pack or fanny pack, will keep you hydrated. Some tour operators and grottos will provide heavy-duty cave packs for carrying water, snacks, etc., but it's best to bring your own pack just in case—be advised, however, that any pack carried underground may never be the same after being dragged through mud and rocks.

Plan on making a complete change of clothes immediately after the trip, including shoes, socks, and underwear. Bring along large plastic trash bags to store your wet, muddy cave clothes so you don't permanently trash your car. Follow the lead of your guide when hiking over private property, dealing with farm and cave gates, and visiting cave owners. Show respect for the cave and your fellow explorers, and you will be in for a memorable and perhaps life-changing experience. Many world-class caving careers have begun with a single grotto novice trip, as have many lifetime friendships.

"Cavers are certainly grounded individuals (no pun intended)," a documentary filmmaker wrote me recently after his first trip underground: "You guys have a quality which makes spending time together in pressurized circumstances actually fun."

THE NATIONAL SPELEOLOGICAL SOCIETY—
2813 CAVE AVENUE
HUNTSVILLE, AL 35810-4431
(256)852-1300
EMAIL: NSS@CAVES.ORG
WEBSITE: WWW.CAVES.ORG

Formed in 1941, The National Speleological Society (NSS) is the world's largest organization concerned with the conservation, scientific study, and exploration of caves. The NSS was formed by a group of cavers in 1941 to promote interest in the study, science, exploration, and protection of caves and their natural contents. Affiliated with the American Association for the Advancement of Science, the NSS has more than 12,000 members throughout the United States and in forty countries.

In many parts of the United States, individual members of NSS have formed local grottos or chapters and regional organizations. These groups sponsor trips, offer training, teach and practice cave conservation, and generally provide a framework for studying caves. A complete list of local grottos can be found on the NSS website at http://www.caves.org. NSS Conservation Policy: (Refer to Act 50-102)

CAVE CONSERVATION POLICY

The National Speleological Society believes: That caves have unique scientific, recreational, and scenic values; that these values are endangered by both carelessness and intentional vandalism; that these values, once gone, can not be recovered; and that the responsibility for protecting caves must be assumed by those who study and enjoy them.

Accordingly, the intention of the Society is to work for the preservation of caves with a realistic policy supported by effective programs for: the encouragement of self-discipline among cavers; education and research concerning the causes and prevention of cave damage; and special projects, including cooperation with other groups similarly dedicated to the conservation of natural areas. Specifically:

All contents of a cave—formations, life, and loose deposits—are significant for its enjoyment and interpretation. Therefore, caving parties should leave a cave as they find it. They should provide means for the removal of waste; limit marking to a few small and removable signs as are needed for surveys; and, especially, excercise extreme care not to accidentally break or soil formations, disturb life-forms, or unnecessarily increase the number of disfiguring paths through an area.

Scientific collection is professional, selective, and minimal. The collecting of mineral or biological material for display purposes, including previously broken or dead specimens, is never justified as it encourages others to collect and destroys the interest of the cave.

The Society encourages projects such as: establishing cave preserves; placing entrance gates where appropriate; opposing the sale of speleothems; supporting effective protective measures; cleaning and restoring over-used caves; cooperating with private cave owners by providing knowledge about their cave and assisting them in protecting their cave and property from damage during cave visits; and encouraging commercial cave owners to make use of their opportunity to aid the public in understanding caves and the importance of their conservation.

Where there is reason to believe that publication of cave locations will lead to vandalism before adequate protection can be established, the Society will oppose publication.

It is the duty of every Society member to take personal responsibility for spreading a consciousness of the cave conservation problem to each potential user of caves. Without this, the beauty and value of our caves will not long remain with us.

Cave Protection Laws In The U.S.

Every state in America has at least one cave and the National Speleological Society estimates there are more than 50,000 caves nationwide. Caves are protected on federal lands by the Federal Cave Protection Act and in 25 states by state cave laws. These laws make it illegal to damage caves or to remove anything found in a cave. To find out if your state has a cave protection law, go to: http//www.caves.org/section/ccms/bat2k /index.htm

You Can Help Save Caves

The National Speleological Society relies on donations to continue the scientific study and preservation of caves. You can contribute to national cave conservation by mailing donations, earmarked for the Save the Caves Fund, to the National Speleological Society (address on opposite page).

The Southeastern Cave Conservancy, Inc. (SCCI) is an organization dedicated to regional cave conservation, caver education, and cave management. It was formed by a group of southeastern cavers in 1991, and currently has about 500 members. Membership is open to all persons interested in caves and cave conservation. For more information, visit www.scci.org.

WORLD'S TOP 10 LONGEST CAVES

NO.	CAVE NAME	STATE	COUNTY	COUNTRY	DEPTH METERS	LENGTH METERS	DATE
1	Mammoth Cave Systam	Kentucky	Ed./Hart/Bar.	USA	571,317	115.5	10/98
2	Optimisticeskaja (Gypsum)	Ukrainskaja	Ternopol	RUSSIA	212,000	15.0	10/99
3	Jewel Cave	South Dakota	Custer	USA	195,615	186.2	11/99
4	Holloch	Schwyz	Muotatal	SWITZERLAND	182,540	941.0	12/99
5	Lechuguilla Cave	New Mexico	Eddy	USA	170,269	77.9	12/99
6	Wind Cave	South Dakota	Custer	USA	150,345	202.4	11/00
7	Siebenhengste-hohgant Hohlensystem	Bern	Eriz/Beat./Ha1	SWITZERLAND	45,000	1340.0	12/99
8	Fisher Ridge Cave System	Kentucky	Hart	USA	144,841	108.6	10/99
9	Ozernaja	Ukrainskaja	Ternopal	RUSSIA	117,000	8.0	10/99
10	Gua Air jernih-Lubang Batau Padeng	Sarawak	Mulu	MALAYSIA	109,000	355.1	03/96

WORLD'S TOP 10 DEEPEST CAVES

NO.	CAVE NAME	STATE	COUNTY	COUNTRY	DEPTH METERS	LENGTH METERS	DATE
1	Lamprechtsofen-Vogelshacht	Salzburg	Leo.Steinberge	AUSTRIA	1632.0	44,000	9/98
2	Gouffre Mirolda / Lucien Bouclier	Haute-Savoie	Samoens	FRANCE	1610.0	9,379	02/98
3	Reseau Jean Bernard	Haute-Savoie	Samoens	FRANCE	1602.0	20,000	Sum90
4	Torca del Cerro (del Cuevon)	Asturias	Picos de Europ	SPAIN	1589.0	2,685	Sum99
5	Shakta Vjacheslav Pantjukhina	Abkhazia	Bol'soj Kavkaz	GEORGIA	1508.0	5,530	1989
6	Sistema Huautla	Oaxaca	Huautla de Ji.	MEXICO	1475.0	55,953	04/99
7	Sistema del Trave (La Laureola)	Asturias	Cabrales	SPAIN	1441.0	9,167	05/97
8	Boj-Bulok	Uzbekistan	Gissarsko-Ala.	UZBEKISTAN	1415.0	14,270	02/97
9	(Il)laminako Aterneko Leizea (BU56)	Nararra	Isaba	SPAIN	1408.0	14,500	Sum88
10	Sustav Lukina jama - Trojama (Manual II)	Velebit	Sjeverni	CROATIA	1393.0	0	05/97

RECOMMENDED READING

Books

Amy, Penny S., and Dana L. Haldeman, eds., 1995. *The Microbiology of the Terrestrial Deep Subsurface.* New York: Lewis.

Borden, James D. and Roger W. Brucker, 2000. *Beyond Mammoth Cave: A Tale of Obsession in the World's Longest Cave.* Southern Illinois University Press, Carbondale, Illinois.

Broad, William J., 1997. *The Universe Below: Discovering the Secrets of the Deep Sea.* New York: Simon & Schuster.

Brock, Thomas D., 1970. *Biology of Microorganisms.* Englewood Cliffs, N.J.: Prentice-Hall.

Brucker, R.W., and Watson, R.A., 1987. *The Longest Cave.* Southern Illinois University Press, Carbondale, Illinois. Knopf, New York. Often cited as one of the best, most exciting books on cave exploration—a true adventure of exploring and mapping Mammoth Cave. A sequel, *Beyond Mammoth Cave: A Tale of Obsession in the World's Longest Cave* will be published in November 2000.

Burgess, Robert F., 1999. *The Cave Divers.* Aqua-Quest Publications, Locust Valley, NY.

Ciba Foundation Symposium 202, 1996. *Evolution of Hydrothermal Ecosystems on Earth (and Mars?).* John Wiley & Sons, New York.

Courbon, P., Chabert, C., Bosted, P., and Lindsley, K., 1989. *Atlas of the Great Caves of the World.* Cave Books, St. Louis.

Dasher, George R.,1997. *On Station.* National Speleological Society, Hunstville, Alabama.

DeKruif, Paul. 1926. *Microbe Hunters.* New York: Harcourt, New York.

Dellenbaugh, Frederick S., 1984 (reprint edition). *A Canyon Voyage: Narrative of the Second Powell Expedition Down the Green-Colorado River from Wyoming, and the Explorations on Land, in the Years 1871 and 1872*, University of Arizona Press.

Dixon, Bernard. 1994. *Power Unseen: How Microbes Rule the World.* Freeman, New York.

Exley, Sheck, 1986. *Basic Cave Diving: A Blueprint for Survival.* National Speleological Society, Hunstville, Alabama.

Exley, Sheck, 1994. *Caverns Measureless to Man.* Cave Books, St. Louis.

Farr, Martyn, 1991. *The Darkness Beckons: The History and Development of Cave Diving.* Diadem Books, London.

Freuchen, Peter, 1995 (reprint edition). *Arctic Adventure.* AMS Press. New York.

Friedmann, E.I., ed. 1993. *Antarctic Microbiology.* John Wiley & Sons, New York.

Gerrard, Steve, 2000. *The Cenotes of the Riviera Maya.*

Gillieson, David, 1996. *Caves: Processes, Development, Management.* Blackwell Publishers, Oxford, U.K.

Gold, Thomas, 1998. *The Deep, Hot Biosphere.* Springer-Verlag, New York.

Gould, Stephen Jay, 1997. *Full House: The Spread of Excellence from Plato to Darwin.* Reprinted. New York: Random House.

Gross, Michael, 1998. *Life on the Edge: Amazing Creatures Thriving in Extreme Environments.* Perseus Press, Cambridge, Mass.

Heslop, Linda, 1996. *The Art of Caving.* New York.

Howes, Chris, 1997. *Images Below: A Manual of Underground and Flash Photography.* National Speleological Society, Hunstville, Alabama.

Klimchouk, A. B., Ford, D.C., Palmer, A.N., and Dreybrodt, W., eds., 2000. *Speleogenesis: The Evolution of Karst Aquifers.* National Speleological Society, Hunstville, Alabama.

Kushner, D.J., ed., 1978. *Microbial Life in Extreme Environments.* New York: Academic Press.

Maurer, Richard, 1999. *The Wild Colorado: The True Adventures of Fred Dellenbaugh, Age 17, on the Second Powell Expedition into the Grand Canyon.* Crown, New York.

McClurg, D.R., 1996. *Adventure of Caving.* D&J Press, Carlsbad, New Mexico.

Moore, G.W., and Sullivan, G.N., 1997. *Speleology: Caves and the Cave Environment.* Cave Books, St. Louis, Missouri.

Northup, D. E., Mobley, E.D., Ingham, K.L., and Mixon, W.W., 1998. *A Guide to the Speleological Literature of the English Language, 1794-1996.* Cave Books, St. Louis, Missouri.

Padgett, Allen, and Bruce Smith, 1997. *On Rope.* National Speleological Society, Hunstville, Alabama.

Postgate, John, 1995. *The Outer Reaches of Life.* Cambridge University Press. Cambridge, U.K.

Powell, John Wesley, and Wallace Earle Stegner, Wallace Earle (Introduction), 1997. *The Exploration of the Colorado River and Its Canyons.* Penguin USA, New York.

Rea, T.G., ed., 1992. *Caving Basics: A Comprehensive Guide for Beginning Cavers.* National Speleological Society, Huntsville, Alabama.

Reames, Stephen, 1999. *Deep Secrets: The Discovery and Exploration of Lechuguilla Cave.* Cave Books, St. Louis, Missouri.

Rhodes, Richard, 1997. *Deadly Feasts: Tracking the Secrets of a Terrifying New Plague.* New York: Simon & Schuster.

Schopf, J.William, 1983. *Earth's Earliest Biosphere: Its Origin and Evolution.* Princeton University Press, Princeton, New Jersey.

Stephens, John Lloyd, Karl Ackerman (Editor) and Frederick Catherwood, 1996 (reprint edition). *Incidents of Travel in Yucatan.* Smithsonian Institution Press, Washington, D.C.

Stephens, John Lloyd, 1969 (reprint edition). *Incidents of Travel in Central America, Chiapas and Yucatan.* Dover Publications, New York.

Taylor, Michael Ray, 1996. *Cave Passages: Roaming the Underground Wilderness.* Scribner, New York.

Taylor, Michael Ray, 1999. *Dark Life.* Simon & Schuster, New York.

Tortora, Gerard J., Berdell R. Funke, and Christine L. Case, 1995. *Microbiology: An Introduction.* 5th (Fifth edition) Redwood City: Benjamin/Cummings.

Weigel, J. and M.W.W. Adams, eds., 1998. *Thermophiles: The Keys to Molecular Evolution and the Origin of Life?* Taylor and Francis, Inc., Philadelphia.

Widmer, Urs, ed., 1998. *Lechuguilla: Jewel of the Underground* (2nd Edition). Speleo Projects, Basel, Switzerland.

You can find more books on cave topics for all ages at:
www.caves.org/service/bookstore
www.speleobook.com
http://cave.divetx.org/book_reviews.html

Publications

The NSS publishes a number of cave-related magazines: The *NSS News* monthly, The *Journal of Cave & Karst Studies* quarterly, and *American Caving Accidents* annually as part of their membership benefits. Members are also invited to purchase the *Speleo Digest*, an annual compilation of stories and articles published in grotto newsletters from around the United States.

RECOMMENDED WEBSITES
Prepared by Carol Zokaites, National Coordinator of Project Underground, Inc.

Caves are very fragile, geologic features that need to be protected for future generations. For this reason, the location of most caves is highly guarded. Viewing a web page is a fun way to experience these caves.

Show caves and National lands caves are managed to allow visitors. These caves offer a wonderful opportunity to experience the underground world in a safe and friendly manner.

The following links were active at press time.

ORGANIZATION LINKS

National Speleological Society (NSS)
www.caves.org/

Show caves of the United States
National Caves Association
www.cavern.com/

CAVES AND KARST PARKS IN U.S.

National Parks
All of the U.S. National Park Service caves, including many lava caves.
www2.nature.nps.gov/grd/tour/caves.htm

Carlsbad National Park
www.nps.gov/cave/

Jewel Cave National Monument
www.nps.gov/jeca/

Mammoth Cave National Park
www.nps.gov/maca/

Wind Cave National Park
www.nps.gov/wica/

MAJOR CAVES OF THE WORLD

Show caves of the World
www.bubis.com/showcaves/index.html
Underwater Caves of Quintana Roo, Mexico
www.primenet.com/~trog/yuccave.html
NSS GEO2 Committee on Long and Deep Caves
www.pipeline.com/~caverbob/
Lamprechtsofen–the deepest cave in the world
http://panda.bg.univ.gda.pl/~dbart/rekord_e.html
Lots of web pages from around the world, divided by country and region.
http://hum.amu.edu.pl/~sgp/spec/links.html

VIRTUAL CAVES

The Virtual Cave
www.goodearthgraphics.com/virtcave.html
Carlsbad Park/tour of Lechuguilla
www.extremescience.com/Lechuguillainfo.htm

Cave Diver Training and Guided Tours
www.cave-dwellers.com
Bat Conservation International
www.batcon.org/

GLOSSARIES

Biospeleology: The Biology of Caves, Karst, and Groundwater
http://www.utexas.edu/depts/tnhc/.www/biospeleology/

Glossary of Cave and Karst Terms:
http://werple.net.au/~gah/speleology/ glossary.htm

Glossary of Caving Terms:
http://wasg.iinet.net.au/glossary.html

Glossary of Speleological and Caving Terms:
http://werple.net.au/~gnb/caving/glossary/index.html#cont

Gander Academy's Caves Theme Page
http://www.stemnet.nf.ca/CITE/cave.htm

PHOTOGRAPHY AND ILLUSTRATIONS CREDITS

Abbreviations for terms appearing below:
(t)-top; (c)-center; (b)-bottom;
MFF-MacGillivray Freeman Films
NGP-National Geographic Photographer

Cover: Stephen L. Alvarez

Table of Contents: 4, Carsten Peter, 5 (t) & (c), MFF; (b), Harris Photographic.

Introduction: Stephen L. Alvarez.

ICE
12-13, MFF
14-19 (all), Carsten Peter
20, John Møller © Arktisk Institut
21, Hazel A. Barton
22, Arktisk Institut
23, Reprinted with the permission of Don Congdon Associates, Inc. © 1935 by Peter Freuchen
25, Chris Blum
26 & 27, Galen Rowell/ Mountain Light
28-29, Jean Pragen/stone
30, Chris Blum
31, MFF
32, Carsten Peter
34 (t) & (b), MFF
35, Brad Ohlund
36-37, Carsten Peter
38, Dr. Colleen M. Cavanaugh, Harvard University
39, Richard B. Hoover/NASA/Marshall Space Flight Center
42-43, MFF
44 & 45, Carsten Peter
46 & 47, Chris Blum
48-49, MFF
50, Martin Mach, Munich Germany (MartinMach@compuserve.com)
51, Dr. Charles Elzinga, Michigan State University
52-54 (all), MFF
56-57 & 58, Carsten Peter
59, MFF
60, Carsten Peter
61, MFF
62, Chris Blum
65 & 66-67, MFF.

WATER
68-69, MFF
70-71, Demetrio Carrasco/stone
72-73, MFF
73, Kenneth Garrett
74, Bill Hatcher
74-75, Kenneth Garrett
76, Frederick Catherwood, courtesy Edizioni White Star
77, MFF
78, Bill Hatcher
82 & 83, Frederick Catherwood, courtesy Edizioni White Star
84, Kenneth Garrett

87, Frederick Catherwood, courtesy Edizioni White Star
88-89, Bill Hatcher
90 & 91, MFF
92, Bill Hatcher
93, Howard Hall/HowardHall.com
94-95, Bill Hatcher
96-97, MFF
100 & 102, Bill Hatcher
103, MFF
104 & 105, Bill Hatcher
106-107, MFF
110, USDCT/Barbara Anne am Ende
111, Wes Skiles/Karst Productions
113 & 114, MFF
116-117 & 118, Bill Hatcher
119-123 (all), MFF
124, Louise D. Hose & David Lester
125, Dr. James Pisarowicz
127 & 128, Bill Hatcher
130, MFF
131 & 132-133, Bill Hatcher.

EARTH
134-135, Michael Nichols, NGP
136-139 (all), MFF
140-141, Stephen L. Alvarez;
142, Bureau of American Ethnology
143, John Burcham
144-145, Jack Dykinga
147, John Burcham
148, Courtesy The John Wesley Powell Museum, Page, Arizona
149, From *Beyond the Hundredth Meridian* by Wallace Stegner; copyright © 1953, 1954 by Wallace Stegner; reprinted by permission of Brandt and Brandt Literary Agents, Inc.
151, Michael Nichols/NGS Image Collection
154-155, Michael Nichols, NGP
156, John Burcham
157, MFF
158 & 159, Stephen L. Alvarez
160 & 162-163, Alan Cressler
165, Stephen L. Alvarez
167, Brent T. Aulenbach
168, Alan Cressler
171, Dave Bunnell
172-173 (all), Michael Nichols, NGP
174-175, Alan Cressler
176 & 177, Harris Photographic
179, Jim Olsen
181, Harris Photographic
182-186 (all), Eric Lars Bakke
187, Michael Nichols, NGP
188-190 (all), Eric Lars Bakke
191, MFF
192 & 193, John Burcham
194-195 & 196, MFF
197, John Burcham
198 (t) & (b), Dr. Merlin D. Tuttle/Bat Conservation International
199, MFF
200-203 (all), John Burcham.

Afterword: Carsten Peter.

ACKNOWLEDGMENTS

AUTHOR ACKNOWLEDGMENTS

Thanks to Tim Cahill and agent Barbara Lowenstein for first bringing *Journey Into Amazing Caves* to my attention, and to my own agent Esther Newberg for her insightful advice on the project.

I owe a great deal to the creative artists of MacGillivray Freeman Films, not only for their unflagging support in allowing me to chronicle the story of their wonderful cave film, but for their willingness to listen to cavers throughout the community and to change the film to reflect the concerns for safety, conservation, and science that are the hallmarks of American caving. While it was always a pleasure working with everyone at MFF, I'm especially indebted to Lori Rick, without whose long hours and attention to detail this book could not possibly exist, and to Steve Judson, who, despite a daunting schedule as the film's director, frequently took time to talk over the book (and to listen as well, always keeping the concerns of cavers foremost in his mind).

I doff my cave helmet to Nancy Holler Aulenbach and Hazel Barton, for letting a nosy stranger pry into their lives, their homes, and their workplaces, always with kindness and grace. I'm grateful to all of the sidebar authors for cheerfully applying their expertise, often on short notice; thanks to you, the book took on meaning and depth that would have otherwise been sorely missing. And a special thanks to sidebar authors Wes Skiles and Hazel Barton; their excellent respective pieces on filming in the Yucatán and juggling the needs of the caver with the needs of the scientist were cut late in the editing process due to space problems; parts of their message survive within the main text, which was surely improved by them. I'm grateful to my old friend (and the perhaps best friend of caves and caving in the world today) Ronal Kerbo for his kind and thoughtful Foreword. Thanks to Jim and Gail Wilbanks for letting me sleep in their comfortable basement and letting me put two publicity-shy cavers in print for the world to see. Likewise, thanks to Donald Davis for permission to quote from his least serious cave theory and for the most serious moral guidance he provides the caving community. Although he appears nowhere in these pages, this book owes a great deal to author and publisher Richard "Red" Watson, who set the artistic standard to which any caver who writes aspires.

At National Geographic Books, I thank editor Kevin Mulroy, along with Johnna Rizzo and Melissa Ryan, for their guidance, professionalism, and patience throughout the long assembly of this book. Thanks also to outside editor Patrice Silverstein and to my reader Lea Ann Alexander for their sharp eyes and keen suggestions.

Thanks to Carol Underwood and Randy and Lisa Duncan for letting me write parts of this story in their houses while they were out of town (well, I did feed the animals), and thanks to Sean Wilson and the Ottawa International Writers Festival for providing the wonderful corner suite in which I wrote the final chapter.

I'm thankful to my colleagues and students at Henderson State University for their patience when I'm off gallivanting underground somewhere. Even greater is the patience shown by my wonderful wife Kathy and my sons Alex, Ken, and Chris, who are always waiting for me on the surface whenever I crawl out of caves or book manuscripts: It's to you that I owe the greatest thanks of all.

Michael Ray Taylor

CONTRIBUTING WRITERS

Barbara am Ende, Ph.D., received a doctorate in geology from the University of North Carolina at Chapel Hill. She first started caving in Iowa in 1973 and continued in various western states, volunteering at Carlsbad Caverns in the 1980s. She helped dig open the entrance to Lechuguilla Cave in New Mexico. In the 1990s am Ende participated in several caving expeditions to Mexico, including an exploration of the Huautla Cave System where she set a new cave depth record in the Americas of -1475 m. Collaborating with the late Dr. Fred Wefer, she created the first fully 3D interactive cave map from data gathered by the Wakulla 2 Expedition in Florida. She currently works at the National Institute of Standards and Technology where she creates scientific visualizations using computer graphics to help scientists better understand their data.

Nancy Aulenbach is a National Cave Rescue Instructor and a former Director of the National Speleological Society. She is instrumental in cave rescues due to her small stature and ability to squeeze through tight passages. She also enjoys a wide reputation for her surveying and exploration skills. In September of 1999, she had the honor of being inducted into The Explorers Club. Like many cavers, Nancy participates in cave conservation efforts (removal of graffiti and garbage from caves), biological inventories, and geology and hydrology studies related to cave formation. An avid caver since she was a child, she currently works as a Montessori School Teacher Assistant in Georgia, where she goes caving nearly every weekend.

Hazel Barton, Ph.D., received her doctorate in microbiology at the University of Colorado Health Sciences Center. She currently works with Dr. Norman Pace at the University of Colorado in Boulder conducting research on drug-resistant tuberculosis. Prior to this, she was an Instructor in the Department of Surgery at the University of Colorado Health Sciences Center. Barton is respected as one of the top cave-cartographers in the country, and has received numerous awards for the maps she has produced of both dry and underwater caves. Hazel is presently a Director and Fellow of the National Speleological Society and the Quintana Roo Speleological Survey.

Louise D. Hose, Ph.D., works as an Assistant Professor of Geology and Environmental Sciences at Chapman University in Orange, California. A geologist and speleologist, she is a Fellow of both the National Speleological Society and The Explorers Club. She also edits the scholarly, multi-disciplinary publication, *Journal of Cave and Karst Studies.* Her scientific investigations of Cueva de Villa Luz and other caves in southern Mexico have been sponsored by the National

Geographic Society's Committee for Research and Exploration.

Ronal C. Kerbo is the national cave management coordinator for the National Park Service, stationed in Denver, Colorado. He has been a cave specialist for the National Park Service for 25 years, and for 13 of those years had the only such position in the Park Service. He is the author or co-author of a number of books, articles, and papers on caves, bats, and diving. He has been caving for over 35 years, and is an Honorary Life Member and a Fellow of the National Speleological Society, a Fellow of the Cave Research Foundation, an Honorary Director of the American Cave Conservation Association, a director of the Karst Waters Institute, and a member or honorary member of many other speleological associations. He is active in caving and cave management in many countries, including China, Australia, South Africa, and Ukraine. He has been profiled in numerous magazine articles and books about caves and environmental issues. He has appeared on national televisions news programs and in television programs about caves.

Jean Krejca is a Ph.D. candidate at the University of Texas at Austin researching cave biology. She also works for the U.S. Fish and Wildlife Service as a karst invertebrate specialist, focusing on preserving caves and the rare and endangered species that inhabit them. For ten years she has been an active member of several speleological associations, she has published results of biospeleological research in peer-reviewed journals and received awards for the work from the National Speleological Society, Cave Research Foundation, Phi Kappa Phi honor society, University of Texas Institute for Latin American Studies and Zoology Department. Her passion for cave exploration takes her on vacations to remote destinations with extended wilderness stays to allow penetration into large tracts of roadless land and the caves within. Some examples include Mexico, Central and South America, and Southeast Asia. Her favorite trips are with a team of good cavers to caves that have a special kind of challenge to enter them, such as ropework, extended underground camps, or sometimes even the use of scuba to continue exploration.

Arthur N. Palmer, Ph.D., teaches Hydrology, Geochemistry, and Geophysics at the State University of New York, College at Oneonta, where he is SUNY Distinguished Teaching Professor and Director of Water Resources Program. He also teaches a summer field course in karst geology at Mammoth Cave National Park for Western Kentucky University. He is an honorary life member of the National Speleological Society and recipient of the NSS Lifetime Achievement Award in Science. He is a fellow of Geological Society of America and the 1994 recipient of GSA's Kirk Bryan Award for his article entitled "Origin and Morphology of Limestone Caves." He is the author of several books on caves, such as *A Geological Guide to Mammoth Cave National Park*, as well as several dozen articles on cave origin. He and his wife Peggy have studied caves and their geology throughout the United States and 13 other countries.

Norman Pace, Ph.D., is Professor of Molecular, Cellular and Developmental Biology at the University of Colorado, Boulder. He has held faculty positions at several institutions, including the National Jewish Hospital and Research Center, the University of Colorado Medical Center, Indiana University, and the University of California, Berkeley. Pace is both a molecular biologist and a microbial ecologist. His laboratory has made substantive contributions in nucleic acid structure and biochemistry and has led the field in the development and use of molecular tools to study microbial ecosystems. This work has led to the discovery of many novel organisms and has substantially expanded the known diversity of microbial life in the environment. Current efforts range from high-temperature environments and microbial ecosystems involved in bioremediation to human disease. He is a member of the National Academy of Sciences; and a Fellow of the American Association for the Advancement of Science, the American Academy of Microbiology, and the American Academy of Arts and Sciences. He has received a number of awards, such as the Procter and Gamble Award in Applied and Environmental Microbiology from the American Society for Microbiology.

Pace has been involved for many years in exploring, mapping, and studying caves. He has led and participated in numerous expeditions in this country

and internationally. Pace has been elected a Fellow of both the National Speleological Society and the Explorers Club, and he received the L. Bicking Award from the NSS for his contributions to American caving.

Charles Shaw, Ph.D., received a doctorate in geology in 1967 at Brown University. He has been a field geologist with the U.S.Geological Survey, taught geology at Windham College, a small liberal arts college in Putney, Vermont, and was a consultant in Houston, Texas for several years. In 1988 he moved to the Mexican Caribbean region south of Cancun with his wife, Kathryn Robinhawk, where they both work for Centro Ecológico Akumal, a non-government organization, working to save the endangered Mesoamerican Reef System. Shaw conducts research into the movement of groundwater through the cave systems of the Yucatán Peninsula with the objective of identifying sources of contamination that endanger the offshore reefs. His research is part of the broader educational and conservation mission of Centro Ecológico Akumal in cooperation with local schools and governmental agencies.

Merlin Tuttle, Ph.D., is an ecologist, award-winning wildlife photographer, and leading conservationist who has studied bats and championed their preservation for more than 40 years. He is currently the Executive Director of Bat Conservation International (BCI), an organization he founded in 1982 that is devoted to research, education, and the conservation of bats. Tuttle is known worldwide through his many media appearances, popular articles, and photographs, including expositions from Harvard University to the British Museum and feature articles in the *Wall Street Journal, New Yorker,* and NATIONAL GEOGRAPHIC. His most recent NATIONAL GEOGRAPHIC article, titled "Saving North America's Beleaguered Bats," appeared in the August 1995 issue. In 1986 Tuttle received the Gerrit S. Miller, Jr. Award, the highest international honor conferred by colleagues in the field of bat biology, and, in 1997 he received both the National Fish and Wildlife Foundation's Chuck Yeager Award and Chevron/ Times Mirror Magazines Conservation Award. In 1999 he was the grand prize winner in Mexico's First Annual Nature Photography Contest—United for Conservation Award.

CREDITS

Published by the National Geographic Society
John M. Fahey, Jr., *President and*
Chief Executive Officer
Gilbert M. Grosvenor, *Chairman of the Board*
Nina D. Hoffman, *Executive Vice President*
President, Books and School Publishing

Prepared by the Book Division
William R. Gray, *Vice President and Director*
Charles Kogod, *Assistant Director*
Barbara A. Payne, *Editorial Director and Managing Editor*
Marianne Koszorus, *Design Director*

Staff for this Book
Kevin Mulroy, *Project Editor*
Patrice Silverstein, *Text Editor*
Melissa G. Ryan, *Illustrations Editor*
Johnna M. Rizzo, *Assistant Editor*
Carl Mehler, *Director of Maps*
Greg Ugiansky, *Map Production*
Marianne Koszorus, *Art Director*
Carol Farrar Norton, *Designer*
Melissa Farris, *Design Assistant*
Michele T. Callaghan, *Copyeditor*
Gary Colbert, *Production Director*
Lewis Bassford, *Production Project Manager*
Meredith Wilcox, *Illustrations Assistant*
Peggy Candore, *Assistant to the Director*
Connie Binder, *Indexer*

Manufacturing and Quality Control
George V. White, *Director*
John T. Dunn, *Associate Director*
Vince Ryan, *Manager*
Phillip L. Schlosser, *Financial Analyst*

CIP DATA

Library of Congress Cataloging-in-Publication Data
   Taylor, Michael Ray, 1959-
      Caves: exploring hidden realms / Michael Ray Taylor.
         p. cm.
         Includes bibliographical references and index.
      ISBN 0-7922-7904-2
         1. Caving. 2. Cave diving. I. Title.

GV200.62 .T38 2001
796.52'5—dc21   00-052710

MacGillivray Freeman's *Journey Into Amazing Caves*
A Film for IMAX® Theatres
Produced by MacGillivray Freeman Films,
    Laguna Beach, California
Produced in association with
    Cincinnati Museum Center with cooperation by
    Ft. Worth Museum of Science & History and
    Milwaukee Public Museum
Major funding provided by
    the National Science Foundation, Michael Cudahy
    and The Endeavors Group.

Production Team
Steve Judson, *Director, Producer, and Editor*
Greg MacGillivray, *Producer and Cinematographer*
Alec Lorimore, *Producer*
Jack Stephens, *Script Writer*
Brad Ohlund, *Director of Photography*
Tom Cowan, *Segment Director*
Wes Skiles, *Underwater Director and Cinematographer*
Gordon Brown, *Ice Cave Cinematographer*
Howard Hall, *Underwater Cinematographer*
Chris Andrei, *Production Manager*
Len Bucko, *Production Manager*
Joshua Colover, *Production Manager*
Arturo Del Rio, *Production Manager*
David Frost, *Production Manager*
Earl Wiggins, *Production Manager*
Robert Walker, *Associate Editor*

Book Production
Lori Rick, *Project Manager*
Matthew Muller, *Image Reproduction Supervisor*

MacGillivray Freeman Team Members
Eric Anderson, Bill Bennett, Chris Blum, Alice Casbara,
Grace Chen, Patty Collins, Mike Clark, Janna Emmel,
Teresa Ferreira, Debbie Fogel, Courteney Hall, Bob Harman,
Kaveh Heravi, Jennifer Karadizian, Dee Kelly, Mike Kirsch,
Mike Lutz, Pat McBurney, Rachel Parker, Ken Richards,
Harrison Smith, Tori Stokes, Kaeran Sudmalis, Susan Wilson

Advisors to the Film
Karol Bartlett, Children's Museum of Indianapolis
Dr. Hazel Barton, University of Colorado, Boulder
Kim Cunningham, Geo-Microbial Technologies, Inc.
Dave Duszynski, Cincinnati Museum Center
Dr. Larry Mallory, Biomes, Inc.
Dr. Luc Moreau, Glaciologie Alpine
National Speleological Society
Dr. Norman Pace, University of Colorado, Boulder
Dr. Arthur Palmer, State University of New York,
    College at Oneonta (also science consultant for book)
Margaret Palmer, Consulting Geologist
John Scheltens, National Speleological Society
Dr. Jill Yager, Antioch College